Psalm 131
The Humility Psalm
O Lord, my heart is not lifted up, my eyes are not raised too high; I do not occupy myself with things too great and too marvelous for me. But I have calmed and quieted my soul, like a child quieted at its mother's breast; like a child that is quieted is my soul.

Humility is one basis of Irish Psychology/Irish Psychiatry. I found the reference to it in Joan Chittister's, The Rule of Benedict: Insights for the Ages. It was selected in eternal, loving memory of my mother, Patricia Coyne, from whom this work originated and to whom this work is dedicated. She quieted my soul.

This book would not have been possible without the Mental Health legislation sponsored by President John F. Kennedy.

C Copyright 1997 by the Hellenist International Institute Publishing Co.

The Hellenist International Institute Publishing Company
A Royalist/Radical Publishing Company
Riverside Plaza M3410
1615 South Fourth Street
Minneapolis, Minnesota 55454

O'Dougherty, Patrick Aquinas, Ph.D., mock Nobel Laureate. 1946--

Irish Psychology/Irish Psychiatry: Catholic Personalist, Intuitionist and Revolution as Therapy--The Tiresias Complex
    Subtitle: A Critique of Medical Revisionism
        Letters Out of Schizophrenia--a Case Study

A Spiritual Dedication to Francis Duff and the Legion of Mary and to the Oblates of St. Benedict
An Intellectual Dedication
    To Mike Franey, David Noble, Betty Agee, Gratia Lee, Mark McGee, Brother Gregory Conant, Father Hilary Freeman, Sister Mary Anthony Wagner, and the Oblates of St. Benedict at St. Benedict College, St. Joseph, Minnesota and St. John's University Collegeville, Minnesota.
Family and Friends
    To Richard and Patricia O'Dougherty-Kast. To my father Aquinas O'Dougherty, my brother Mike and my sisters: Margaret Mary, Mary Ann and Maureen. To Megan, Sean and Stephen. To John O'Dougherty, John Herlick, the Conant family and the Linstroth family.

ISBN: 0-9626665-8-0

Library of Congress Catalog Card Number: 97-93445
Printed in the United States of America

The Crucible of Irish Psychology/Irish Psychiatry--A Subordinated Epistemology
Viktor Frankl Experiences Out of Schizophrenia

The writer, Patrick O'Dougherty, first became a psyche consumer at St. John's University, Collegeville, as an undergraduate in 1964. After transferring to the University of Minnesota in 1965, he began seeing a clinical psychologist there named Dr. Shirley Corrigan. Taking a break from graduate school at the University of Minnesota, he spent a month at Anoka State Hospital and then hitchhiked down to Atlanta, Georgia. This city was chosen because it is the black capitol of the South and produced Martin Luther King. For artistic definition, the writer precipitated an act of psychological civil disobedience across the street from Martin Luther King's Ebenezer Baptist Church and was arrested. Following several other minor arrests he was sentenced to Bellwood Chain Gang for eight months. Later on he spent time in Woodside Receiving Hospital in Youngstown, Ohio, eight months in Massillon State Hospital in Massillon, Ohio. He has also been a psyche consumer at Hope Transition Center, the Old Miller Hospital, and United Hospital all of which are located in St. Paul. He lived for seven years at Guild Hall which is located in the F. Scott Fitzgerald neighborhood of St. Paul. In Minneapolis he has been a psyche consumer at Fairview/St. Mary's Hospital and at Hennepin County Medical Center. The counseling people who had the greatest personal impact upon my artistic journey were Dr. Shirley Corrigan at the University of Minnesota, Dr. Thomas Graham at Massillon State Hospital, and Dr. Jane Rozsnafszky of Minneapolis. Dr. Jane Rozsnafszky is quite brilliant. Deedee Daniels at United Hospital was quite empathic. Dr. Irving Bernstein, psychiatrist, of Minneapolis got me my scholarship. Dr. Patrick Stokes and Dr. Joseph Gendron, psychiatrists, had a powerful mentoring influence on me. Joyce Forsgren, Neale Thompson and Bill Caldwell of Hennepin County Medical Center were excellent therapists. Father Michael Tegeder of the Cathedral Parish in St. Paul got me involved with the Benedictine Oblates. He is the St. Benedict of the psyche consumers in the Twin Cities area. I especially want to highlight the contribution to my psychological life made by my psychologist sister, Dr. Margaret Wright, and her husband, Dr. Frank Wight, pediatric neurologist. I never thought I would write a book in medicine, however, this work came out in response to their personal and medical contributions.

St. Dymphna, Dorothy Day, Cardinal Newman, and St. Patrick

The religious figures who had the greatest influence on me are St. Dymphna, the Irish patron saint of the mentally ill, who has a shrine at Massillon State Hospital in Massillon, Ohio, Dorothy Day, Cardinal John Henry Newman, and St. Patrick. I have been involved on and off with the Dorothy Day Movement in Minneapolis and St. Paul for twenty-two years. For example, I worked for six and one half years in the Dorothy Day room at Newman Center building an agriculture research library for the International Alliance for Sustainable Agriculture. I have also worked in and frequented the Dorothy Day Centers in the Twin Cities. A major part of the experiences of the work came out of the Dorothy Day Movement, my seven year involvement at Guild Hall in St. Paul with the residents and staff there, especially, Jan and Russ Divergsten, and my thirty year on and off relationship with Newman Center at the University of Minnesota. Mike Franey, my physicist friend, and Mark McGee, a psychologist and friend, love these psyche experiences. Newman is my guiding light. St. Patrick is my psychological mentor. Like St. Patrick, I'm "up from slavery."

TABLE OF CONTENTS

IRISH PSYCHOLOGY/IRISH PSYCHIATRY: THE TIRESIAS COMPLEX

IRISH PSYCHOLOGY/IRISH PSYCHIATRY
Ourselves for self-gift, An Irish Psyche Consumer Motto, a schizophrenic dialectic with "Ourselves Alone," the Sinn Fein Motto

STATEMENTS OF INTERPRETATION AND INTENT AND DEFINITIONS AND ASSUMPTIONS
CATHOLIC
RELIGIOUS LIBERTY: THE INTERPRETATIVE BASIS OF IRISH PSYCHOLOGY/IRISH PSYCHIATRY

Religious freedom or liberty is the basis of Catholicism and Irish Psychology/Irish Psychiatry. This theme highlights the lack of freedom found in many Catholic countries, for example, in the Third World. Freedom or liberty shows "the absence of a bond or limitation."[1] Thus it is largely nonmaterialistic. Materialistic relations have an element of determinism in them. Determinism is a denial of freedom or of free will.[2] The basis of Irish Psychology/Irish Psychiatry is Catholic freedom or liberty. It is largely nondeterministic. By the nature of limitation Catholic freedom differentiates three kinds of liberty. First, there is physical liberty where one is free from physical constraints. Second, there is psychological liberty. This is synonymous with free will. Third, there is moral liberty which is the absence of a moral obligation forbidding an act. To be free, an act must contain all three levels of freedom. For example, I am free to go to Ireland if I have a passport and fare (physical liberty); if I can decide to go (psychological liberty); and if I do not have a duty or obligation to remain at home to protect my children (moral liberty). One may be physically and psychologically free to commit suicide but one can never be morally free to commit suicide because the Fifth Commandment prohibits it.[3]

Religious Freedom: Physical Liberty
Private Life/Public Life
How does one apply these three distinctions to religious freedom? First, in private life a person has the right to freedom from restraint about religion (a physical liberty).[4] Coercion used to force an unbeliever such as a pagan or a Jew to believe is an evil. The faithful have the right to constrain unbelievers to prevent them from putting obstacles in the way of faith by blasphemy, evil persuasion or persecution. A person's private worship shall not be infringed upon.[5] With respect to physical freedom, public authorities frequently intervened in the private religious devotion of individuals in France during the Revolution and in Soviet Russia. Martin Luther King, Roisin McAliskey and Ghandi all worked for physical freedom.[6]

In public life can one be coerced or forced to act? The Church teaches one can use no direct constraint to force a person to belong to the Catholic Church.[7] The Catholic Church thinks false religions have no right to public manifestation. The state can, however, either forbid or tolerate them.[8] Moreover, in public life one does not

1

have a right to a "public profession of error" because tolerance of error is provisional.[9]

Religious Freedom: Psychological Liberty

Second, man can only believe through free will (psychological liberty).[10] Materialism does not determine free will. Atheism and agnosticism are basically deterministic positions. Furthermore, no human authority, no State, no community of States authorize teaching that which is contrary to moral good or religious truth. Moreover, "failure to prevent this with civil laws and coercive measures can be justified in the interests of a higher and more general good."[11]

Religious Freedom: Moral Liberty

Third, moral liberty relates to faith. The true religion, the Catholic Church, is the basis of moral liberty. Many people have adhered to false religions; but, this does not make them subjectively guilty. They do not, however, have the moral liberty to do this.[12] One is free to serve God as God expects service. It is a human right. Before the state, the freedom to practice religion founds itself in God and bases itself on his revelation. On the right of moral liberty, "Error has no rights." A person cannot claim a right or a virtue through their error.[13]

Moral liberty applies to the just and true. Thus a person does not have the moral liberty to belong and spread a false religion. A person cannot claim any special right by error. Again, "Error has no rights."[14]

The Church and Liberty

The Church has physical, psychological and moral liberty. First, the Church makes her decisions without coercion. Second, the Pope and the College of Cardinals exercise free will. Third, the Church is the bastion of faith and morals which define moral liberty.

Church and State

Ultimately, Christian liberty is witness to the absolute and just dominion of God over man, and to the duty of the person toward God.[15] The State has no competence in religious matters. It should not intervene in the religious domain except when public order is at risk.[16] By setting up "'the freedom of all religions without distinction, truth is confused with error.'"[17] Civil society must unite to the religious society of the Church and subordinate itself to it.[18] St. Thomas Becket, for example, was martyred for "his defense of Church liberty from the power of the state."[19]

2

IRISH PSYCHOLOGY/IRISH PSYCHIATRY: THE TIRESIAS COMPLEX

The Liberal Conception of Liberty

The liberal makes liberty an end in itself.[20] For example, the liberal conception of "freedom of worship" guarantees all religions the alleged right to public exercise and equal protection.[21] "Catholic liberty becomes divine by submitting to God; liberal liberty destroys itself by trying to make itself God."[22] The Vatican council declares that the human person has a right to religious freedom in all three of its dimensions: physical, psychological and moral. This freedom consists in immunity from coercion by individuals, social groups and human powers.[23] In public life Pope Leo XIII said, no one shall be forced to embrace the Catholic faith against his will.[24]

Catholic Freedom and Irish Psychology/Irish Psychiatry

In order for psychology and psychiatry to be _free_, the human area must contain all three dimensions of liberty: physical, psychological and moral. Any limitation on any of these areas much have definition. Behaviorism, for example, is often a denial of all three dimensions of liberty. First, subjects often lack physical liberty or are put in constraints. Second, B.F. Skinner denied free will or psychological liberty. Third, Skinner thought behaviorism was "beyond freedom and human dignity."[25] He denied the Church's conception of moral liberty which has a basis in human dignity.

Human Dignity: A Basis of Irish Psychology/Irish Psychiatry

What is the basis of human dignity? Dignity is the superiority possessed by a being as a result of its perfection. It reflects the fullness of its qualities about other beings of like kind or of a different kind. Supreme dignity belongs to God only and is his majesty. The person is the most perfect in nature and possesses the highest dignity in creation. Because they lack reason, animals are "individuals" not "persons." Intelligence transcendentally focuses on truth. The perfection of the person one finds in knowledge of and adherence to truth. Error does not confer dignity. Adherence to error by a person is a forfeiting of his or her dignity. Truth is the foundation of liberty. The Bible says, "The truth will set you free" and "The truth will deliver you."[26] Freedom from constraint is not the primary basis of liberty or the basis of the dignity of the person. Truth rather than liberty is the prior foundation of human dignity. Freedom to be wrong decreases human dignity. The use of coercion to stop the public practice of a false religion is an offense against human dignity. The Church or State should not prevent a person from obeying his or her conscience, especially about religious matters.[27] Conscience is the foundation of human dignity.

Government Psychology/Psychiatry: The Divine Model

IRISH PSYCHOLOGY/IRISH PSYCHIATRY: THE TIRESIAS COMPLEX

St. Thomas said "the governments of men must take as their model the divine government from which they derive."[28] He further said, "Civil society, although distinct from the religious society of the Church, must be united to it and subordinated to it."[29] Jesus Christ imposes the supernatural order not only on individuals and families, but on societies too. Cardinal Pie said, "'Perfect harmony between priesthood and empire is the common law and normal condition of Christian societies.'"[30] The State should serve God's majesty.[31] State psychologists/psychiatrists should have a divine model and serve God's majesty.

The State and the Common Good

The Church thinks (1)"What does not conform to truth and to moral law has objectively no right to exist, and right to propaganda and action. (2) the fact of not preventing it by State legislation and coercive measures can nevertheless be justified for the sake of a superior and wider good."[32] The Church "does not forbid authority to tolerate what is at variance with truth and justice, for the sake of avoiding some greater evil, or of obtaining or preserving some greater good."[33] For the common good, human law may tolerate evil, "it may not and should not approve of or desire evil for its own sake."[34]

"Common Right": The American Way

"Common right" is the idea that some freedom should be given all religions and all sects. This is the situation in America. Leo XIII argued in his encyclical, Longinqua oceani, that in America the Church would produce "'even more fruits if she enjoyed, not just freedom, but a favoured treatment by the laws and the protection of the public authorities.'"[35] The public exercise of a person's own peculiar worship propagates indifferentism.[36]

Moral and Physical Liberty

The problem of religious liberty revolves largely on the distinction between moral and physical liberty. Moral liberty in religious matters can belong only to the true religion. No one has the moral liberty to belong to Islam or Buddhism. Physical liberty is immunity from constraints. A person is physically free to belong to Islam or Buddhism. In private life immunity from constraints either coercive or prohibitory is a right. In public life there must be no forced adherence to Catholicism--no coercive restraints. However, there is no right to immunity from this sort of constraint.[37]

The Resolution of Catholic Freedom in Christ

4

IRISH PSYCHOLOGY/IRISH PSYCHIATRY: THE TIRESIAS COMPLEX

Catholic freedom finds its resolution in Christ. It requires a total dedication to Catholic sacrifice, love, service, order, civilization, theology, goodness, truth, and beauty to fulfill this resolution. All are one in Christ.

Irish Psychology/Irish Psychiatry and Freedom

Irish Psychology/Irish Psychiatry is truly free because it contains all three types of liberty: physical, psychological, and moral. It has a spiritual basis. Irish Psychology/Irish/Psychiatry has a divine model. It bases itself on human dignity and emphasizes the primacy of conscience. It resolves the problems of immunity from constraints in private life and in public life maintains there should not be forced adherence to Catholicism. It favors rights and responsibilities for professionals and practitioners. It finds its fulfillment and its praxes in the idea that freedom is breaking the bonds of colonialism. Freedom is anti-colonialism.

IRISH
Principles of Irish Psychology/Irish Psychiatry

The basic principles of Irish Psychology/Irish Psychiatry are, first, unconditional love, especially Christian love. Second, the mediatrix of women, for example, the Blessed Mother and Mary Magdalene in therapeutic relationships is crucial. Mary and Mary Magdalene were important figures at the crucifixion. Of the apostles, only John was at the crucifixion. Thus women, for example Mary and Mary Magdalene, must be an integral dimension in any form of therapy. Third, the idea that the sexes are radically equal. Fourth, the idea that the races are radically equal. For example, the Irish, the French, the Germans, the Indians and the Polynesians are radically equal. Fifth, the idea that the many cultures in the world are radically different, like potatoes and oranges.[38] Sixth, personalist intuitionism is the basis of Irish Psychology/Irish Psychiatry. It counters the depersonalization of the Irish white Negro experience and the depersonalization many women, minorities and gays have experienced throughout history. Seventh, a Catholic Thoreau experiment in civil disobedience is the basis of therapy. Psychiatric commitment, for example, is a recapitulation of the Thoreau experiment in civil disobedience. Eighth, psychology and psychiatry are theological Catholic sciences. Ninth, asceticism is a therapeutic basis of psychology and psychiatry. Clients and professionals should not be "slaves to passion."[39] Tenth, religious freedom is the basis of the sciences of psychology and psychiatry. For example, the minimum acceptable dosages of psychiatric drugs is necessary to enhance the personal freedom of the client. Eleventh, therapy is cause related. For example, an acceptable outcome of therapy is involvement in a cause and the radicalization of the person. Revolutionary thinking, for example, in a field, like ecology, is a desired outcome of therapy. The twelfth principle of Irish

IRISH PSYCHOLOGY/IRISH PSYCHIATRY: THE TIRESIAS COMPLEX

Psychology/Irish Psychiatry is the principle of stew. The Irish are famous for their stew. However, they do not want to turn their race into stew. Thus the perfection of race, especially in Christ, is a principle of Irish Psychology/Irish Psychiatry.

THE IRISH WHITE NEGRO THEME
Discrimination/Depersonalization

The Irish, woman and many minorities, many indigenous minorities, and many psyche consumers have had the status of white Negroes. **Psyche consumers are like white Negroes because they often face discrimination, stereotyping, and oppression. When one group is diminished all groups are diminished.**

'"We viewed ourselves as Ulster's white Negroes--a repressed and forgotten dispossessed tribe captured within a bigoted, partitionist statelet that no Irish elector had cast a vote to create...'"[40] Finnbarr O'Doherty

Irish discrimination has universal implications, specifically, spiritual, political, and intuitive implications for psyche consumers. Christ faced discrimination. Christ opposed discrimination.

Rights/Responsibilities

ulster's white negroes, is an important work which looks at how a civil rights and civil liberties struggle in Ireland changed into a revolutionary war. This war claimed 3,000 lives in Northern Ireland and raged for a quarter of a century or more. ulster's white negroes, details the beginning of the civil rights and responsibilities movement in Ireland. It depicts the wave of working class Catholics, in Derry, who moved to throw off the yoke of second class citizenship. This book documents this movement's confrontation with the state, the 1969 introduction of British troops, and the Bloody Sunday massacre of thirteen unarmed protestors. Finnbarr O'Doherty, the author of ulster's white negroes, was an activist in the Derry Housing Action Committee. He was a co-founder of the Northern Ireland Civil Rights Association. O'Doherty is an integral part of this struggle from the elections to the Battle of Bogside.[41]

In short, the Irish faced discrimination and depersonalization. The following is an example of the discrimination against and depersonalization of the Irish and of all people:

The Irish White Negro Status: The Special Powers Act

The Special Powers Act was retained by the Six Counties in Northern Ireland. It applied to no other region in the United Kingdom. This Special Powers Act has been continuously in operation since 1922.

6

It empowered authorities to:
"(1) Arrest without warrant.
(2) Imprison without charge or trial and deny recourse to Habeas Corpus or a court of law.
(3) Enter and search houses without warrant, and with force, at any hour day or night.
(4) Declare a curfew and prohibit meetings, assemblies (including fairs and markets) and processions.
(5) Permit punishment by flogging ('cat-of-nine-tails').
(6) Deny any claim to a trial by jury.
(7) Arrest persons that police desired to examine as witnesses, forcibly detain them and compel them to answer questions, under penalties, even if answers might incriminate them. Such a person was guilty of an offence if he/she refused to be sworn or answer a question.
(8) Do any act involving interference with the rights of private property.
(9) Prevent access of relatives or legal advisers to a person imprisoned without trial.
(10) Prohibit the holding of an inquest after a prisoner's death.
(11) Arrest a person who 'by word of mouth' spreads false reports or makes false statements.
(12) Prohibit the circulation of any newspaper, e.g. the United Irishman etc.
(13) Prohibit the possession of any film, or gramophone record.
(14) Arrest a person who does anything "calculated to be prejudicial to the preservation of peace or maintenance of order in Northern Ireland and not specifically provided for in the regulations."
(15) The Act allows the Minister of Home Affairs to create new crimes by government decree, e.g., it became a crime to name a club a 'Republican Club.'"[42]

A Philosophic Response to Discrimination and Depersonalization: Personalism
Personalism is a philosophy which counters and heals the depersonalization many psyche consumers, women, minorities, gays and men have experienced. Personalism: "any doctrine or movement which emphasizes the rights and centrality of the individual being in his social, political intellectual, ... milieu."[43]   Pope John Paul II says, **"A person is an objective entity, which as a definite subject has the closest contacts with the whole (external) world and is most intimately involved with it precisely because of its inwardness, its interior life."**[44] For example, the Legion of Mary reflects the social, political, and intellectual dimensions of the person in their external and interior life. Catholic personalism is the philosophy of Pope John Paul II.

Love and the Person

7

IRISH PSYCHOLOGY/IRISH PSYCHIATRY: THE TIRESIAS COMPLEX

The word 'love' has many meanings. Wojtyla starts from the perspective "that love is always a mutual relationship between persons."[45] This relationship bases itself on attitudes to the individual and the collective good. The relationship should be reciprocal. Love is more than an attraction. It builds on the good. It knows values. Love is desire. It is not reducible to desire alone. Love is also goodwill or selflessness. Love exists between a man and a woman. Reciprocity contains reliability and durability. Love structures itself by interpersonal communion. Love transforms sympathy into friendship. Love realizes itself in relationships. It unifies persons. Engaged love Wojtyla argues is "the surrender of one's 'I.'"[46]

Personalism and Mathematics

A Definition of Mathematics
Mathematics is a language, especially of science. A language is "a special set of symbols, letters, numerals, rules, etc. used for the transformation of information, as in a computer."[47] A science is "a branch of knowledge or study esp. one concerned with establishing and systematizing facts, principles, and methods as by experiments and hypotheses." In particular mathematics deals with "the group of sciences (including arithmetic, geometry, algebra, calculus, etc.) dealing with quantities, magnitudes, and forms, and their relationships, attributes, etc., by the use of numbers and symbols"[48] However, mathematicians cannot agree on a definition of the field. It has a spiritual, personal, undefined and intuitive feminine dimension.[49] **Intuitionists, like Jan Brouwer, think mathematics is activity.** For example, it is a pregnant methodology. A methodology is "the science of method, or orderly arrangement; specif., the branch of logic concerned with the application of the principles of reasoning to scientific and philosophical inquiry."[50] **A methodology is also an activity, like pregnancy.** Also, evolution is a necessary element in any definition of mathematics. Is personalism mathematics? Yes, mathematics applies as the language and methodology of a philosophy which defines the person, for example, an evolving woman.

The Intuitionist School of Mathematics

How do mathematicians define the intuitionist school? The key is the question of existence. Jan Brouwer disagreed with Poincare on mathematical existence. Brouwer did not think mathematical existence meant freedom from contradiction as Poincare thought but **intuitive constructibility**. Brouwer conceived of mathematics as a **free ("mind")** building mathematical objects beginning from **self-evident primitive intuition.** The psychological/psychiatric free "minds" should redefine primitive intuition. Do psychological/psychiatric thoughts have a self-evident basis? Yes, it is found in the activity of the mind and will. In contrast, formal logic is a means of describing regularities in systems already built. It lacks value in proving the

8

foundations of mathematics. The absolute validity of logical principles is questionable. The axiomatic foundation of mathematics failed. Brouwer intended that David Hilbert would be unable to prove the consistency of arithmetic. Even if Hilbert succeeded, Brouwer thought this would insure a mathematical system defined by axioms.[51] **The mind of psyche consumer is a free ("mind") building from primitive objects of intuition."[52]** The Catholic conception of freedom which contains physical, psychological and moral components which define freedom in the mathematical context as well. Primitive mathematical objects of intuition must have physical, psychological and moral dimensions to be truly free. The free mind has a moral dimension, for example.

How is an intuitive theology expressed? It is personal. Psychological intuition and the intuitionist school of mathematics have a theological base expressed in Christ. For example, Christ faced discrimination, had Jewish roots, and dealt with rights and responsibilities issues. He dedicated his life to sacrificial love and was a virgin. Christ faced an unequal situation in the Roman Empire. Christ had the Beatific Vision, a vision that transcends the infinite, always before him. It _is_ personal and intuitive. Was Christ emotionally unstable? My father, Aquinas O'Dougherty, thought either he was God or he was insane. Christian beauty has a projective intuitive dimension. So does Christian intelligence. All these themes come together in the intuitive theology of love and freedom. **Love and freedom are intuitionism.**

Intuitionism and Personalism: The Mathematical Integration of the Person

Intuitionism or activity is part of the definition of the person because activity is part of the objective as well as interior life of the person. Could the person also be an incomplete definition for the field of mathematics? The answer is yes. The person has a **free** ("mind") building from primitive objects of intuition."[53]

DOROTHY DAY/PERSONALIST

One example of a personalist revolutionist is Dorothy Day who was a personalist. She faced depersonalization by an abortion experience. Dorothy Day resolved her problems with loneliness and depersonalization in the birth of her daughter, her conversion and in community, specifically the Catholic Worker community she founded.

First, Another Kind of Revolution

Dorothy Day and Glory's Season

9

IRISH PSYCHOLOGY/IRISH PSYCHIATRY: THE TIRESIAS COMPLEX

Dorothy writes on a fast that she went on in Rome on October 1st. during the last session of the Council. She and twenty other women were fasting.[54] Chanterelle del Vasto started the fast. The women recited each morning after Mass the Our Father, St Francis's peace prayer and the Eight Beatitudes. Following these prayers, there was communal singing. Then the women retired to their rooms.[55]

For each day the women had a schedule. Mass was at seven-fifteen; and, after the Mass was communal prayer. From nine to twelve was silence or reading. A vigil in the chapel continued throughout the day. At noon was communal reading. The readings were from Martin Luther King; and, a description of the life of Father Paul Gauthier who started the Order of Companions of Jesus the Carpenter in Nazareth.[56]

Members of the Ark, a religious group, gave a presentation on nonviolence. During the afternoon lectures were given by priests; and, at six a Doctor gave support. The group prayed again at seven in the evening. Dorothy offered her fast for famine victims. Her bed was hard. After the fast, her physical pains left her.[57]

St. Therese/Personalist
the "Little Way"

Dorothy Day writes of St. Therese of Lisieux. She says that Therese "speaks to our condition. Is the atom a small thing? And yet what havoc it has wrought. Is her little way a small contribution to the life of the spirit? It has all the power of the spirit of Christianity behind it. It is an explosive force that can transform our lives and the life of the whole world once put into effect."[58] Dorothy says that Therese, a "saint, of this day, is releasing a force, a spiritual force, upon the world" to fight nuclear incineration.[59] Nonviolence, as Dorothy interpreted it, is not separate from Biblical nonviolence and from "the little way" of St. Therese.[60]

Another Catholic Worker, Martin Rodgers/Personalist

Martin is a twenty-four year old African American Catholic. He says that the Church teaches us that in Christ, we are all brothers and sisters. Martin says that in Africa or India people greet each other with the phase, Aman, which means "I recognize God in you."[61] He says that at the Last Judgement God tells us that he will judge us on what we do to the least of our brethren. He contends, "Racism, sexism, heterosexism, age-ism, classicism, and, indeed militarism" block our ability to view each other as brothers and sisters.[62]

Martin points out that in the Catholic Church in America nearly 50% of the members are non-white; and, world wide it approaches 75% non-white. We are all

IRISH PSYCHOLOGY/IRISH PSYCHIATRY: THE TIRESIAS COMPLEX

God's children. We are all equal. We should all see God in each other.[63] The philosophy of the Catholic Worker is Christian personalism.

Personalism and the Irish Cause

ulster's white negroes represents personalism in this philosophy's fullest dimension. This Irish movement emphasizes the person's rights and personal centrality in his or her social, political, intellectual environment.[64] This work in Irish Psychology/Irish Psychiatry comes out of the Irish experience and gives it a Christian, personalist, intuitionist face.

SCIENTIST: ST. THOMAS AQUINAS

St. Thomas Aquinas thought "Sacred doctrine is a science."[65] He contends "we must bear in mind that there are two kinds of sciences. There are some which proceed from principles known by the natural light of the intellect, such as arithmetic and geometry and the like. There are also some which proceed from principles known by the light of a higher science: thus the science of optics proceeds from principles established by geometry, and music from principles established by arithmetic. So it is that sacred doctrine is a science because it proceeds from principles made known by the light of a higher science, namely, the science of God and the blessed. Hence, just as music accepts on authority the principles taught by the arithmetician, so sacred science accepts the principles revealed by God."[66]

Sacred doctrine is a higher science. The Eight Beatitudes are an example of the scientific principles made known by the light of this higher science. This is the science of God and the blessed.[67]

Personalism and Science
Love: A Soul Science

Personalist psychology is "the science of the soul" which opens to analysis the structure of man's inner life.[68] **Thus love, an area in psychology, is also a soul science. Algebra is a language of this soul science. They interconnect in personal activities.**

REVOLUTIONIST: THE EIGHT BEATITUDES

Irish Psychology/Irish Psychiatry have many scriptural foundations. The Bible is a revolutionary document as well as a theological document. For example, the Eight Beatitudes are an example of revolutionary statements made in the Bible.

11

IRISH PSYCHOLOGY/IRISH PSYCHIATRY: THE TIRESIAS COMPLEX

The Eight Beatitudes

Matthew 5: 3-11
> 3 "Blessed are the poor in spirit, for theirs is the kingdom of heaven.
> 4 Blessed are those who mourn, for they shall be comforted.
> 5 Blessed are the meek, for they shall inherit the earth.
> 6 Blessed are those who hunger and thirst for righteousness, for they shall be satisfied.
> 7 Blessed are the merciful, for they shall obtain mercy.
> 8 Blessed are the pure in heart, for they shall see God.
> 9 Blessed are the peacemakers, for they shall be called sons of God.
> 10 Blessed are those who are persecuted for righteousness sake, for theirs is the kingdom of heaven."

By incorporating these beatitudes into the Irish Psychology/Irish Psychiatry movements they make these movements more revolutionary than the Communist Manifesto.[69] They are the cornerstone of the of the radical Irish Psychology/Irish Psychiatry rights and responsibilities movements. For example, psyche consumers are poor in spirit, need comfort, are meek, seek satisfaction, need mercy, need pure environments, need peace and suffer persecution.

"Words Are Deeds": An Approach to the Bible, A Dictionary in Miniature and to the Eight Beatitudes

The writer went to a symposium on Wittgenstein where the idea "words are deeds" originated.[70] Mike Franey thinks there is a problem with this idea because of the discontinuity between words and deeds. For example, one may announce the Eight Beatitudes, however, living them out as deeds is difficult. The Eight Beatitudes serve as a preamble to the radical dictionary below. Keeping Wittgenstein's theme "words are deeds" in mind, the above Eight Beatitudes should apply to each of the terms in this dictionary. There must be little discontinuity between the words and the deeds of the radical psychology/psychiatry movements. For example, with the term "psychological activism--lobbying or demonstrating for psyche causes" each of the Eight Beatitudes should apply to this term.[71] For example, this is true for every psyche cause involved in psychological activism. A try is necessary by the radical to have little discontinuity between the words in this dictionary, the Eight Beatitudes and his or her actualizing of these terms. In God the discontinuity between words and deeds resolves. In the science of sacred scripture, God's "words are deeds."

"Words Are Deeds"
Dictionary of Green Radical Psychology/Psychiatry Terms
A Dictionary in Miniature

12

IRISH PSYCHOLOGY/IRISH PSYCHIATRY: THE TIRESIAS COMPLEX

This Dictionary Transposes the Radical Environment's Movement Vocabulary into Psychology and Psychiatry Creating a New Field Irish Ecopsychology/Ecopsychiatry

1. Psychological activism--lobbying or demonstrating for psyche causes[72]
2. Activist psychology/psychiatry--an Irish approach
3. Psychological Adat--Adat "recognizes the rights of all the life-forms in a place, as well as those yet to arrive in the next generation and those that have passed from the scene through death."[73] This is a tribal approach to psychology. It is egalitarian.[74] Psychological Adat is the counseling expression of this approach.
4. Anarcho-psychologists--psychologists opposing psyche hierarchies[75]
5. Antidrug movement--one reaction to psychiatry[76]
6. Psychological anarchism--an emphasis on informal primary groups, that is, a hospital society without hierarchy[77]
7. Psychological antinature--any act that works against the nature of the person.[78] Industrialism is an example. It works against the nature of the person, for example, much of capitalist technological culture.
8. Anti-Animal Farm--the psyche hospital
9. Antipsychological activists--dissidents against professionals
10. Psychological "Aquarius Festival" Recreation Therapy[79]
11. Authoritarian tradition--It is institutional, system centered and powerful.[80]
12. Brain tree spiking--controlling drugs that affect the brain stem[81]
13. Cartesian psychology--the abstract self in an objective world leads to the exploitation of nature[82]
14. Communal Property Rights Amendment--This amendment would counter, Amendment V, which protects private property rights--psyche property is communal is the theme[83]
15. Psychology/psychiatry of confrontation--psychotage is an example--it is confrontation to create change--political confrontation analysis is another example[84]
16. Conservative outlook--constant psychological crises
17. Counterfriction--Patrick O'Dougherty's approach to psychology--"life as counterfriction" is a theme from Thoreau[85]
18. Psychological "cracking"--mental illness; This word counters radical environmental acts like draping pictures of cracks on dams.[86] In psychological cracking, the psyche is like a cracked mirror.
19. Credo of psyche nonviolence--the basis of patient client interaction[87]
20. Cultural extinction--the cultural decline of a civilization[88]
21. Psychological cultural discontent--counterfriction in any cultural setting[89]

IRISH PSYCHOLOGY/IRISH PSYCHIATRY: THE TIRESIAS COMPLEX

22. Cultural suicide--the death of a culture; the Afro-American writer John Edgar Wideman thinks the black people experienced cultural suicide.[90] Scientologists think psychiatry perpetuates this phenomenon with minorities.
23. Psychological death of nature--the death of the earth; geocide[91]
24. Psychological psychiatric decay--personal regression; brain cell decay[92]
25. Psychological deconstruction--pushing the meanings of academic, professional and patients words to extremes[93]
26. Democratic tradition--client centered psychology/psychiatry
27. Detested--beyond test results[94]
28. Psychological/psychiatric discontent--any member of the antipsychiatry movement like many African-American, Chicano, Native American and Irish thinkers.
29. Discontinuity--one basis of psychology and relationships; Psychiatric discharge creates a discontinuity.[95]
30. Psychological "disequilibria"--unequal relationships in therapy; Human waste creation is another example.[96]
31. Distraction theory--the theory of how to distract a client from deviant behavior[97]
32. Psychological dystopia--the pathological relations within an institutional setting or environment[98]
33. Psyche First--the idea professionals have no right to subdue clients with restraints or drugs or through legal commitments
34. Psychology First: Radical Psychological Integrity[99]
35. Psychological integrity of the earth--a holistic approach to the environment[100]
36. Psychological psychiatric elites--the psyche hierarchy[101]
37. Psychological environmentalism--the position of the psychological integrity of all life--the interrelatedness of all life
38. Environmentalist psychology/psychiatry-the idea that psychology is primarily a tribal movement
39. Psychological "egalitarian relationships"--one approach to psyche relationships[102]
40. A Bill of Rights for All Species--the basis of Environmental Psychology
41. Psychological errata--schizophrenic loose thinking in writing[103]
42. Escalations--patient conflicts in hospital medication management are an example; conflicts that result in commitments are another example
43. Evolution--a position in psychology accepted by the Catholic Church--One half of the American people and about one half of the Protestants do not accept this position.
44. Father and Mother figures--the business dimensions of psychology and psychiatry; government psychologists and psychiatrists[104]

IRISH PSYCHOLOGY/IRISH PSYCHIATRY: THE TIRESIAS COMPLEX

45. Psychological Faustianism--the try to seize control of evolution both personal and societal[105]
46. Foucault and psychology--the idea that the "circuity of power" is the main trait of psychology/psychiatry[106]
47. Psychological Gaia--an earth person such as a tribal member[107]
48. Gang of Twelve--the twelve largest psychological organizations in the world[108]
49. Psychological generational conceptions--African Adat in psychology is a good example[109]
50. Golden Rule--the professional with the handbook rules[110]
51. Psychological Green--any inroad to a livable psychological environment
52. Green Wall--the dollar in psychology[111]
53. Psychological Green Rage--psychology plus environmental activism; an example might be opposition to lethal dimension of American capitalist culture[112]
54. Psychological guerrilla theaters--psychologists in war fronts like Vietnam--psyche movies
55. Psychological integrity of human nature--the integration of the person approach in psychology
56. Psychological interdependence--group solidarity like Catholic or Polish solidarity
57. Irish Psychology/Irish Psychiatry--a cause or activist or revolutionary approach to counseling; this approach encourages revolutionary innovative thinking in academic fields
58. Liberal outlook--an emphasis on minimizing psychological crises in patients
59. Lowest Common Denominator in psychology--the committed[113]
60. Lethal culture: any culture that creates pathology. Much American culture is lethal to good psychology. Capitalism is an example.
61. Psychological "limits to growth"--limits of the person[114]
62. Psychological love canal--transference[115]
63. Psychological mediation movements--crisis interventions[116]
64. Psychological Molotov cocktail--explosive personal interactions; an overdose on medications[117]
65. Psychological monkeywrenching--psychological vandalism[118]
66. Monoculture--channeled therapies[119]
67. Psychological monolithic ideologies--behaviorism; Marxist Psychology[120]
68. Mythic discourse--a universal order that enables the dominant psychological professional group to impose its linguistic will on the therapeutic community, for example, psychiatric and psychological diagnosis.[121]
69. Natural law--The Catholic approach to psychology and psychiatry favors laws of nature; death and gravity are natural laws. There are many.

IRISH PSYCHOLOGY/IRISH PSYCHIATRY: THE TIRESIAS COMPLEX

70. Natural law therapy--basic therapeutic opposition to technological culture
71. Negative psychological dictum--"nothing is natural," for example, an over focus on pathology[122]
72. Nonviolence is unnatural--birth, for example, is a violent process
73. Psychological nonviolence--kissing, reassuring, and touching for reinforcement[123]
74. Psychological particularist ideologies--personalism
75. Personalism--a psyche philosophy emphasizing the rights and centrality of the person in all dimensions. Patrick A. O'Dougherty is a Catholic personalist intuitionist. Like the person, the earth has integrity. Intuitionism is the activity as opposed to the logic school of mathematics. Intuitionism, or activity, integrates personalism. For example, activities fine tune the personality.
76. Positive psychological dicta--"Everything is natural."[124] It is an overemphasis on the positive in the person--the Mary Poppins' approach
77. Potomac psychologist--A National Institute of Mental Health Lobbyist is an example. Anyone seeking a government psychology or psychiatry grant.[125]
78. Psychological preroading--initial interview[126]
79. Primeval therapy--primitive family therapy
80. Primitive psychology--characterized by the discontinuity of power and Adat.[127]
81. Primitivism--tribal reactions and approaches to psychology and psychiatry[128]
82. Problem civilization-the pathology is not the person in therapy but industrial civilization[129]
83. Psychocapitalist dominion--the dominion of capitalism over the psyche and mind
84. Psychecultists--New Age Thinkers have a cult dimension; Scientologists[130]
85. Psyche culture--one dimension of all culture
86. Psyche ethic--an action is right when it tends to preserve the integrity, stability and beauty of the psychological community[131]
87. Psychesphere--the brain[132]
88. Psychocentric civil rights--personalist psychology
89. Psychodefense--A Handbook of the Militant Defense of Personal Integrity[133]
90. Psychodefense of consumers--legal defense of psyche consumers[134]
91. Psychofascism--pushing the authoritarian tradition to the limit[135]
92. Psychofascist dominion--Fascist dominion over the psyche;[136] Fascist experiments in psychology and psychiatry; an example is the Fascist psychiatrist Ernst Rudin[137]

93. Psychological civil disobedience--any intentional act against authorities with psychological implications, for example, the Newman Center case of Patrick O'Dougherty

94. Psychological reductionism--an example is reducing psyche research to statistics--statistical reductionism

95. Psychological resistance points--activism[138]

96. Psychological resource protection--personal property

97. Psychological restoration--in history it is any recovery of the past like the recapture of Ancient Israel in the modern time.

98. Psychological retrograde--any personal regression[139]

99. Psychological psychiatric sensibility--the idea that actions fine tune the person[140]

100. Psychological sensibility--the psyche ethic

101. Psychological social construct theorists--"human nature is a societal category."[141]

102. Psychomartyr--anyone willing to die for a psyche cause[142]

103. PsychoMarxist dominion--Marxists dominion over the psyche

104. Psychoracism--the apartheid psychology of the South African leader/psychologist Hendrik Verwoerd;[143] any psychological theory with racist implications, for example, Herbert Spencer's idea "'inferior'" people should die to improve racial stock or the thinking of Francis Galton who created the term "eugenics."[144] Another psychoracist was Benjamin Rush, the "'father' of American psychiatry," who thought that a disease created the skin color of blacks and miscegenation would spread it throughout the white population.[145]

105. Psychosocialists--many American liberal psychologists--European Socialist psychologists--any communal psychologists[146]

106. Psychosocialist dominion--the dominion of socialists over the psyche[147]

107. Psychosystem--any systemic approach to psychology

108. Psychotactics-restraints or legal appeals are examples[148]

109. Psychotage--potentially damaging confrontation to create psychological change; Also, it justified rage in psychiatry and psychology by either professionals or patients.[149]

110. Psychotastrophe--any type of psychological meltdown such as severe depression or problems do to overmedication.[150] Brain cell decay.

111. Psychoteurs--psyche activists; proponents of psychotage[151]

112. Psychotopes--psychologically distinct regions in the brain[152]

113. Psychotechnology--the analysis of the technological domination

114. Psychotopia--a healthy psychological institution or environment[153]

115. Stewardship--the psychiatrists role; the term has a religious connotation; psychology and psychiatry are basically spiritual professions

116. Thoreau experiment in psychological civil disobedience--the commitment process[154]
117. Tiresias Complex--the complex of an individual whose downfall is truth; Tiresias was a Greek prophet and mythological figure whose downfall was truth.
118. Psychological tree sitting--chair sitting in isolation;[155] intellectually or emotionally putting yourself out on a limb
119. Psychological/psychiatric underground--any disaffected psyche consumer who drops out intellectually, politically, socially, or economically.
120. Urban Walden--a psychiatric hospital, for example, Hennepin County Medical Center, Minneapolis.
121. The psychological utopian vision--wilderness is therapy[156]
122. Virgin forest--new experiences
123. Worldview--a universal outlook, for example, a Catholic outlook[157]

This dictionary calls the reader to put these words into deeds. Some of the words in this dictionary, for example, psychofascism, have negative meanings. Give a Catholic response to these words. For example, apply the Eight Beatitudes to psychofascism.

"No Justice, No Peace"
The Irish and Their Cause/
Cause Therapy or Revolutionist Therapy

A strengths of the Irish people is that they have a cause or causes. ulster's white negroes links the causes of the Irish people with the civil rights causes of the Negroes. It links with the struggles of women and children throughout the world. The Irish causes form Christian, universal, social, political, intellectual, intuitive, mathematical, and spiritual Declarations of Interdependence. These Irish causes should link with the causes of all psyche consumers in a new form of therapy, cause therapy, based on the scriptures, for example, the Eight Beatitudes.

Praxes of Cause Therapy: The Green Party

Praxes of cause therapy are the themes of the Green Party: "Ecological Wisdom, Social Justice, Grassroots Democracy, Community-based Economics, Feminism, Respect for Diversity, Personal and Global Responsibility and Future Focus."[158] The Catholic freedom, personal intuitionist and scriptural definition and integration of these themes within the new field of Irish Ecopsychology/Irish Ecopsychiatry, which the radical dictionary partially defines, are the basis of Irish

IRISH PSYCHOLOGY/IRISH PSYCHIATRY: THE TIRESIAS COMPLEX

Ecopsychology/Irish Ecopsychiatry. Irish Psychology/Irish Psychiatry is about everybody. For Irish therapy, look to the cross.

PREFACE TO LETTERS OUT OF SCHIZOPHRENIA

AN ETHNOGRAPHIC CASE STUDY OF PATRICK A. O'DOUGHERTY, PH.D., AN IRISH CATHOLIC

"Mental Illness is Not a Choice"
My Weakness, My Strength[159]
Prayer in Time of Controversy

"Lord Jesus Christ,
I am involved in a bitter controversy
in which it is very difficult to tell who is right and who is wrong.
I cannot understand why these things take place among people who are trying to live Christian lives, and I wish I were not involved in it.
Let me realize that controversy is a fact of life and that even Your life was filled with it.
Help me to accept whatever comes in a spirit of resignation
and teach me to be better from this experience.
Grant that I may resolve the controversy in a Christian manner and lead others to You because of it."[160]

Image Projected/Psychological Reality

Patrick O'Dougherty's projected image contains the elements of Catholic, Irish, white Negro, scientist and revolutionist. The dynamism of his personality is the radical integration in Catholic freedom of his projected image and his psychological reality. The psychological reality reflects in his MMPI profile, his Jungian personality test and in his Strong Vocational Interest Test which follow:

ID: 00074751   RPT DATE: 15-
SEX: Male   EDUC: 20
AGE: 47   MARS: Never M
SETTING: College Counseling

Raw	L	F	K	Hs	D	Hy	Pd	Mf	Pa	Pt	Sc	Ma	Si	A	R	MAC
Score:	1	0	22	0	14	24	14	29	10	5	9	11	12	4	16	14
K Corr.				11			9			22	22	4				
T Scr.	39	36	64	45	40	57	50	56	49	51	58	39	34	42	52	32

?'Cannot Say (Raw): 0

FB (Raw): 0
F-K (Raw): -22   Cooke's Disturbance Index: 445

Welsh Code (new): 835 74/612:90#   K-/:LF#   Percent True :
Welsh Code (old): 5'834-76/129:0#   K-/F?L: Percent False:

PROFILE VALIDITY

This is a valid MMPI-2 profile. However, the client approached the ite
a defensive and overly cautious manner. He may be evasive and unwillin
admit to many personal faults. Individuals with similar profiles are o
uninterested in and unwilling to enter into discussion of their problem

SYMPTOMATIC PATTERNS

This MMPI-2 clinical profile is within normal limits. The client did n
report psychological conflicts or situational stresses that are produci
great difficulty for him at this time. He appears to be dealing effect
with his life situation, and seems to be obtaining sufficient satisfact
out of life at this point.

He appears to be more interested in abstract matters than practical
activities. He has a somewhat unconventional approach to life and may
problems in a creative and unusual way. He may appear dreamy and
preoccupied and may show some dissatisfaction with life. He appears to
no sex-role conflicts.

In addition, the following description is suggested by the content of t
client's responses. The client does not appear to be an overly anxious
person prone to developing unrealistic fears. Any fears he reports are
likely to be viewed by him as reality-based rather than internally
generated. His item content suggests he is rather non-competitive and
passive, generally easygoing, and uncritical in interpersonal situation
He appears to be happy with life and feels positive about the future.
response content reflects a high degree of self-confidence and the abil
to deal with life tasks. He reports that his work situation is general
satisfactory. No significant negative work attitudes requiring treatme
attention were noted in his item content.

INTERPERSONAL RELATIONS

Quite outgoing and sociable, he has a strong need to be around others.
is gregarious and enjoys interpersonal attention.

DIAGNOSTIC CONSIDERATIONS

This profile is within normal limits and no clinical diagnosis is provi

----

Summary of Approaches of Each Archetype*

	Orphan	Martyr	Wanderer	Warrior	Magician
Goal	Safety	Goodness, care, responsibility	Independence, autonomy	Strength, effectiveness	Authenticity, w ness, balance
Worst Fear	Abandonment, exploitation	Selfishness, callousness	Conformity	Weakness, ineffectuality	Uncentered sup ciality, alienati from self, othe
Response to Dragon	Denies it exists or waits for rescue	Appeases or sacrifices self to save others	Flees	Slays	Incorporates an affirms
Spirituality	Wants deity that will rescue and religious counselor for permission	Pleases God by suffering, suffers to help others	Searches for God alone	Evangelizes, converts others, spiritual regimes, disciplines	Celebrates expe of God in ever respects differe ways of experie the sacred
Intellect/ Education	Wants authority to give answers	Learns or forgoes learning to help others	Explores new ideas in own way	Learns through competition, achievement, motivation	Allows curiosit learns in group alone because fun
Relationships	Wants caretaker(s)	Takes care of others, sacrifices	Goes it alone, becomes own person	Changes or molds others to please self, takes on pygmalion projects	Appreciates dif ence, wants pe relationships
Emotions	Out of control or numbed	Negative ones repressed so as not to hurt others	Dealt with alone, stoic	Controlled, repressed to achieve or prevail.	Allowed and le from in self an others
Physical Health	Wants quick fix, immediate gratification	Deprives self, diets, suffers to be beautiful	Distrusts experts, does it alone, alternative healthcare, enjoys isolated sports	Adopts regimes, discipline, enjoys team sports	Allows health, body to exercis good food
Work	Wants an easy life, would rather not work	Sees as hard and unpleasant but necessary, works for others' sake	"I'll do it myself," searches for vocation	Works hard for goal, expects reward	Works at true tions, sees wor its own reward
Material World	Feels poor, wants to win lottery, inherit money	Believes it is more blessed to give than to receive, more virtuous to be poor than rich	Becomes self-made man or woman, may sacrifice money for independence	Works hard to succeed, makes system work for self, prefers to be rich	Feels prosperou with a little or has faith will a have necessitie does not hoard
Task/ Achievement	Overcoming denial, hope, innocence	Ability to care, to give up and give away	Autonomy, identity, vocation	Assertiveness, confidence, courage, respect	Joy, abundance ceptance, faith

*The Innocent is not included on the chart because it is not an heroic archetype. When we live in paradise, there is no need for goals, fears, tasks, wo etc. The Innocent is both pre- and post-heroic.

Self-Test

Discover the Archetypes Dominant in Your Life

Indicate how frequently the following statements reflect your attitudes by scoring them from 0 to 4: Never = 0; Seldom = 1; Sometimes = 2; Frequently = 3; Always = 4. After taking the test, see columns below that categorize the statements into archetypes. Total your score in each category. Nine or more in each suggests that the archetype is active in your life; fifteen or more suggests it is very active.

1. _1_ It's important to be careful. Other people will cheat you when they can.

2. _3_ I find that when I change my attitudes my environment changes.

3. _1_ Most important to me right now are identity issues. I'm not sure who I am.

4. _4_ I push hard to prove myself and to succeed.

5. _2_ The world is good and I am safe and cared for.

6. _2_ I feel very alone, but it gives me satisfaction to see that I can make it on my own.

7. _4_ The most important thing is loving.

8. _1_ I often feel disappointed in or betrayed by other people.

9. _2_ All seeming problems really are illusions. I can assert God's love/the perfection of the universe and once again see that all is well.

10. _4_ I am very competitive and really enjoy winning.

11. _3_ Times have been rough, but I've learned to cope.

12. _2_ I find out about my own shadow self by what upsets me in others.

13. _1_ I use drugs/alcohol to get high and feel better. (Or: I use shopping, work, or frantic activity to divert myself from problems.)

14. _3_ I expect people I meet to be trustworthy.

15. _4_ When challenged, I stand up for myself and, if necessary, fight to defend myself.

16. _4_ I'm in a new job/doing my job differently/undertaking a new course of study.

17. _4_ I expect to be loved and cared for.

18. _3_ I struggle hard for the causes/ideas/values I believe in and against those that are wrong or harmful.

19. _2_ I frequently give people more than I get back.

20. __0__ What I really want is someone to take care of me, but there is no one who will/can really care for me.

21. __2__ When I am betrayed or unjustly treated, it reminds me to take pains to be fair to others.

22. __3__ I love to travel/study/experiment because I find I learn about myself and the world when I do.

23. __2__ I see no evil, hear no evil, speak no evil.

24. __4__ I feel most myself when I'm creating something new.

25. __4__ I want my life to make a difference, to make a mark on the world.

26. __3__ When I stay calm and centered, others seem quieted too.

27. __2__ If others could just see the light, they could have as wonderful a life as I do.

28. __5__ Since I've changed, my world has changed radically. Years ago, I would not have imagined things would turn out so well.

29. __3__ I think I'm justified in feeling superior to other people: I'm smarter, or better educated, or stronger, or more disciplined, or hardworking, or have better values, or because of my sex, my racial or ethnic heritage, my class, my accomplishments, my beliefs.

30. __2__ Tragedies (accidents, illnesses) often happen to me and those around me.

31. __3__ I work hard but do not expect to be rewarded or appreciated adequately for what I do.

32. __2__ If I could only win that jackpot, all my problems would be solved.

33. __4__ I feel good about myself and grateful for my life.

34. __3__ I would like to be more appreciated by others.

35. __3__ I'll do whatever life requires of me. I want to make whatever contribution I can.

36. __3__ I sometimes avoid or sabotage intimacy with others in order to maintain my freedom.

ocent	Orphan	Wanderer	Warrior	Martyr	Magician
2	#1 _1_	#3 _1_	#4 _4_	#7 _4_	#2 _3_
	8 _1_	6 _2_	10 _4_	19 _2_	12 _2_
	13 _1_	11 _3_	15 _4_	21 _2_	24 _4_
	20 _0_	16 _6_	18 _2_	31 _3_	26 _2_
	30 _2_	22 _3_	25 _4_	34 _3_	28 _3_
15	32 _2_	36 _3_	29 _2_	35 _3_	33 _4_
	Total _7_	Total _16_	Total _20_	Total _17_	Total _11_

To see the relative importance of each archetype in your life, chart
your scores on the grid below.

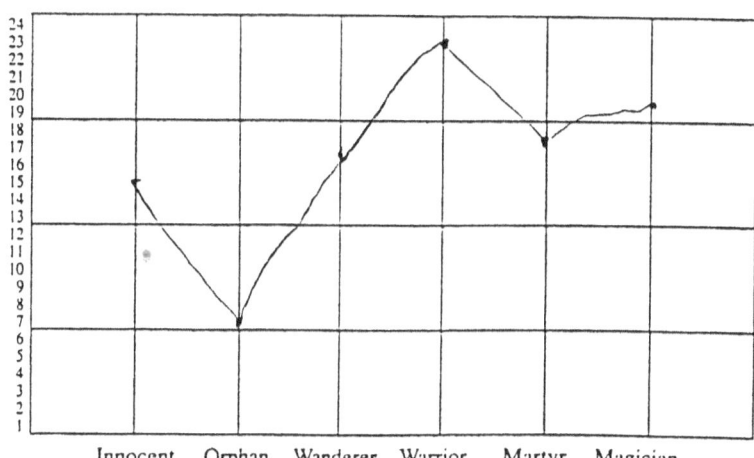

Innocent   Orphan   Wanderer   Warrior   Martyr   Magician

STRONG VOCATIONAL INTEREST TEST-MEN

HANKES REPORT FORM FOR—   SEE OTHER SIDE FOR EXPLAN

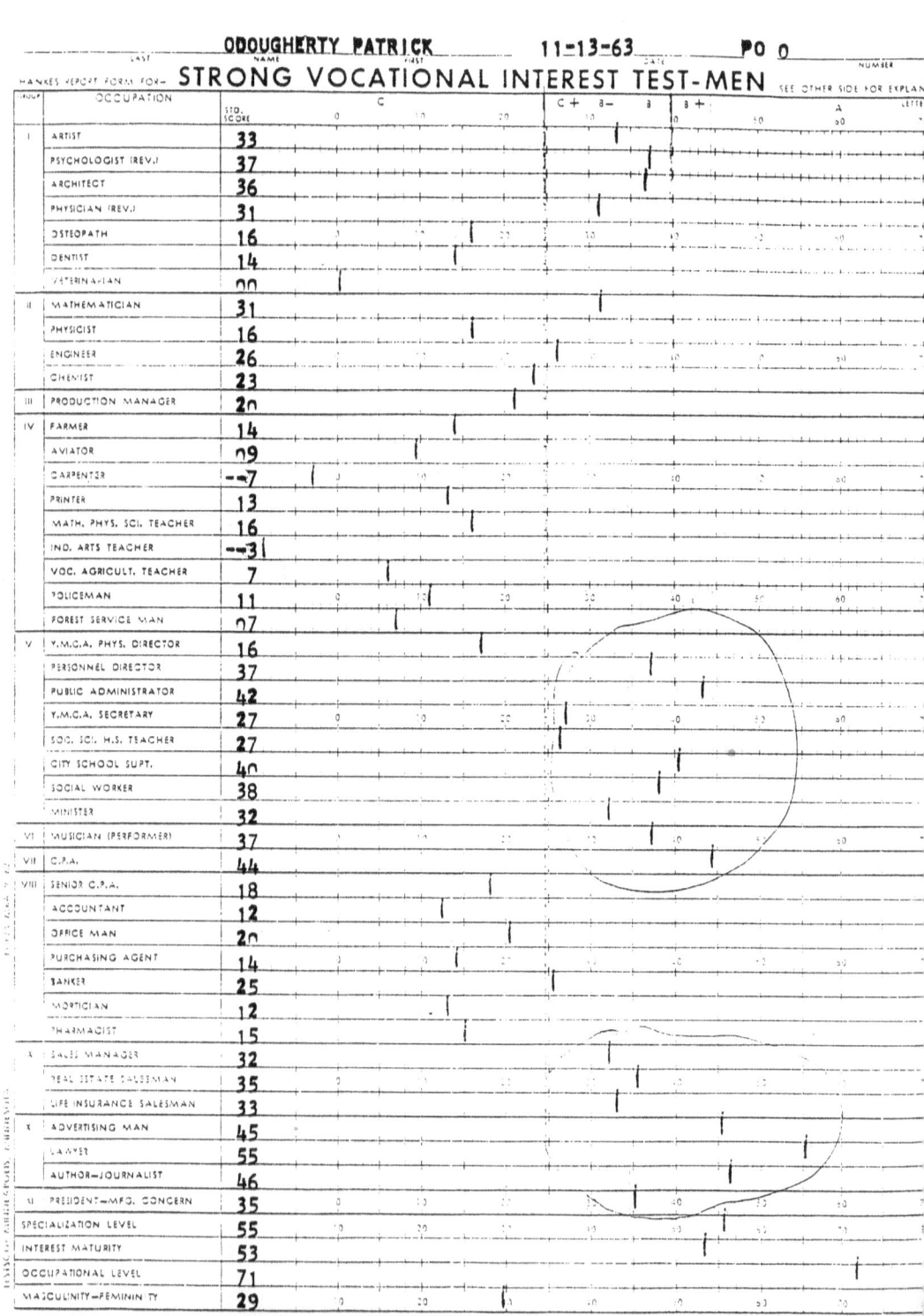

GROUP	OCCUPATION	STD. SCORE
I	ARTIST	33
	PSYCHOLOGIST (REV.)	37
	ARCHITECT	36
	PHYSICIAN (REV.)	31
	OSTEOPATH	16
	DENTIST	14
	VETERINARIAN	00
II	MATHEMATICIAN	31
	PHYSICIST	16
	ENGINEER	26
	CHEMIST	23
III	PRODUCTION MANAGER	20
IV	FARMER	14
	AVIATOR	09
	CARPENTER	--7
	PRINTER	13
	MATH, PHYS, SCI. TEACHER	16
	IND. ARTS TEACHER	--3
	VOC. AGRICULT. TEACHER	7
	POLICEMAN	11
	FOREST SERVICE MAN	07
V	Y.M.C.A. PHYS. DIRECTOR	16
	PERSONNEL DIRECTOR	37
	PUBLIC ADMINISTRATOR	42
	Y.M.C.A. SECRETARY	27
	SOC. SCI. H.S. TEACHER	27
	CITY SCHOOL SUPT.	40
	SOCIAL WORKER	38
	MINISTER	32
VI	MUSICIAN (PERFORMER)	37
VII	C.P.A.	44
VIII	SENIOR C.P.A.	18
	ACCOUNTANT	12
	OFFICE MAN	20
	PURCHASING AGENT	14
	BANKER	25
	MORTICIAN	12
	PHARMACIST	15
IX	SALES MANAGER	32
	REAL ESTATE SALESMAN	35
	LIFE INSURANCE SALESMAN	33
X	ADVERTISING MAN	45
	LAWYER	55
	AUTHOR—JOURNALIST	46
XI	PRESIDENT—MFG. CONCERN	35
	SPECIALIZATION LEVEL	55
	INTEREST MATURITY	53
	OCCUPATIONAL LEVEL	71
	MASCULINITY—FEMININITY	29

r occupational interests are recorded by heavy
in the scales opposite the appropriate occupations.
the example below the man has a B rating in the in-
of artists (note the B at the top of the report

blank), a C rating in the interests of psychologists, an A
rating in the interests of architects, and a B– rating in the
interests of physicians.

OCCUPATION STANDARD SCALE		0			C +	B –	B	B +		A	
		0	10	20	30	40	50	60	70		
	37										
OLOGIST (REV.)	20										
TECT	47										
IAN (REV.)	32										

A rating means that the individual has the inter-
f persons successfully engaged in that occupation; a
ing means that the person does not have such inter-
and the ratings B+, B, and B– mean that the person
-ly has those interests but we cannot be so sure of
act as in the case of A ratings. It is seldom that
s with C ratings are found in the occupation, and if
gaged they are either indifferent successes who are
to drop out or are carrying on the work in some more
s unusual manner. The latter situation is exempli–
y a physician with a rating of C in the interests of a
ian who is engaged as superintendent of a hospital.

.gh ratings B and A should be considered. One
hoose one occupation so rated or plan to utilize
interests in two or more such occupations. Thus, if
ores high in both law and engineering one might
e for both and become a patent attorney, or a law–
ecializing in engineering problems.

. higher a score to the right of the shaded area the
r the certainty that one has the interests character–
f that occupation. The lower the score to the left
shaded area the greater the certainty that one does
ve the interests of the occupation. Scores falling
the shaded area are indeterminate; they help some–
to show, along with other scores, the general trend
's interests in an occupational group. But general y
an be ignored. Consequently, in the above diagram
ores for psychologist and physician are disregarded
e conclude that the individual has an A rating in the
urs of an architect and a B rating in the interests of
s.

ucations included in the same group all correlate
with one another.

n's interests change very little from 25 to 55 years
. They change somewhat from 20 to 25 years and

much more so from 15 to 20 years. Consequently, the
younger the man, particularly below 20 years of age, the
less certainly can his interests be identified in terms of
some occupation. Such changes in interests as take place
are more likely to result in higher ratings than the reverse.
This is particularly true with respect to ratings in Group V.

The ratings from this test should not be viewed as con-
clusive; they are not guaranteed. Instead they should be
viewed as merely suggestive and to be considered in the
light of all other information bearing upon one's vocational
choice. Occupations rated A and B+ should be carefully
considered before definitely deciding against them; occupations rated C, C+, and B– should be carefully considered
before definitely deciding to enter them. Remember only a
few from among all the hundreds of occupations are re-
ported on here.

Remember also this is a test of your interests. Your
abilities must also be considered. Interests point the way
you want to go, abilities determine how well you can pro–
gress.

Scores on the four special scales (see bottom of report
sheet) are for the use of trained counselors and should be
explained personally by them. The IM scale expresses
maturity of interests. One's age must be taken into con–
sideration in interpreting this score. It applies only to men
between the ages of 15 and 20 . The OL scale indicates
whether one's interests are similar to common workmen
(a low score) or to business and professional men (a high
score). The MF scale indicates whether one's interests are
similar to the interests of men or women. The average man
scores 50 on the OL and MF scales. The Specialization
Level Scale pertains to college men as to whether or not
they would enjoy advanced study and narrow specialization
in their work.

See the author's VOCATIONAL INTERESTS of MEN and
WOMEN.

Edward K. Strong, Jr.
Professor of Psychology
Stanford University, California

IRISH PSYCHOLOGY/IRISH PSYCHIATRY: THE TIRESIAS COMPLEX

This approach to his schizophrenia is an <u>art</u> because it has psychological free will. The tests are mathematical and statistical <u>scientific</u> anecdotes or ashes. The <u>theology</u> which emerges is <u>transcendent integration</u>. Integration is a civil rights theme. The idealization and integration of the Catholic, Irish, white Negro, scientist and revolutionist themes occur in Christ and in the scriptures. Idealizing these themes makes them transcendent. **The healing, resolution, transcendence, universalization and integration in a Catholic ethnographic case study, like the one of Patrick A. O'Dougherty, is found in Christ who was a member of the Jewish ethnic group, the chosen people. The Bible is an ethnographic case study. All are one in Christ.**

An Irish Catholic Thoreau Experiment in Psychological Civil Disobedience

Dedicated to Martin Luther King

<u>The writer proclaims unconditional love and forgiveness to everyone mentioned in these letters</u>. He wrote them over six years. The book that came out of these letters came to a head at Newman Center at the University of Minnesota in an incidence of psychological civil disobedience with the campus minister there, Father Steve Bossi. Patrick A. O'Dougherty has been a member of Newman Center on and off since the 1960s. He was always well received there. However, in 1991 there was change of staff and an Italian priest, Father Steve Bossi took charge. <u>His agenda was Italian Catholic Marxism</u> with no dissent. For example, at Newman, they had Brechtian liturgies. He was an East German Marxist dramatist. The Pathfinder Bookstore, a political socialist bookstore/movement, was given a propaganda table at Newman Center. And, a Sandinista leader was given a seminar in the basement of Newman Center. One dimension of Patrick O'Dougherty's agenda was the Legion of Mary. The two males, Patrick O'Dougherty and Father Steve Bossi, locked horns in a friendly dispute. Patrick O'Dougherty got no recognition, no concessions and no rights at Newman Center for five years. He published five books during this time and built a major research library in sustainable agriculture at the International Alliance for Sustainable Agriculture during this period. To resolve this dispute, Patrick O'Dougherty contested the Newman Center budget at the Chancery and contacted a black civil rights attorney about filing a case. The attorney told him to call the City Attorney. The City Attorney directed him to open an MP case on Father Steve Bossi. The case number is MP96092631.

Patrick O'Dougherty then sent a statement and the case number to his friends in the congregation. A minor psychological civil disobedience trespassing incident followed. Father Steve Bossi had Patrick O'Dougherty brought into Hennepin County Crisis Intervention. A court case, a seven month stay of commitment at HCMC which is an urban Walden, and two books, including <u>St. Patrick, The Green</u>

IRISH PSYCHOLOGY/IRISH PSYCHIATRY: THE TIRESIAS COMPLEX

Revolution, and the Hydrogen Conversion Project and this work Irish Psychology/Irish Psychiatry, came out of this Catholic Thoreau case in civil disobedience. Father Steve Bossi did not go to the trial. The case is before God. Both of the priests at Newman Center, Father Steve Bossi and Father Richard Colgan moved on. The Hennepin County Medical Center file for this case is 049-43-16. The petitioner File No. is P2-96-60223 done by Gary Fischler on May 13, 1996 at C-300 Government Center 300 South Sixth Street Minneapolis, MN 55487. Dr. Karen Bruggemeyer diagnosed Patrick O'Dougherty as schizoaffective-bipolar. To address a long-term medication problem created by tardive dyskinesia, Dr. Karen Bruggemeyer prescribed a very small dose of Risperdal for him. Patrick O'Dougherty's MMPI Adult Clinical System Interpretative Report listed his high scales (the Raw Score follows the scale) as K for Defensiveness (22), Hysteria (24), Paranoia (10) and Schizophrenia (9). His low scale scores are Depression (14) and Social Introversion (12).[161] In her evaluation of him Joyce Forsgren his coordinator at HCMC told him that he was "brilliant," but had problems with "looseness of thinking, religiosity, agitation, negativism, proving superior, hypercritical, and imposing on others."[162] For coping strategies, she recommended "taking meds, seeing my psychologist and spirituality."[163] He has a richness of therapy and dialogue with his psychologist, Dr. Jane Rozsnafszky. Patrick O'Dougherty has never heard voices or had any hallucinations. All of his tests came out normal.

A Religious Political Prisoner Out of Newman Center

For just making a minor initiative in a lawsuit at Newman Center, Patrick O'Dougherty got a seven month stay of commitment at HCMC. The length of this commitment made him think he was a political prisoner at HCMC. His case was also systemic. It was the product of capitalist, imperialist, colonialist society. For example, the problem at Newman Center was psychological oppression by an American military man, Father Steve Bossi, of the Catholic, Irish, white Negro, scientist and Catholic revolutionist dissident, Patrick O'Dougherty.

The O'Dougherty Family Structure

Patrick O'Dougherty experienced a critical, negative achieving family. T h e O'Dougherty family is like the family portrayed in Pat Conroy's The Great Santini. It is an Irish Catholic military family. The family had twenty-five to thirty years active in WWII. In contrast to the Great Santini father, the father in the O'Dougherty family was a small town humble, religious, and intelligent Irish Catholic lawyer. In the O'Dougherty family, the woman all have very strong dominating personalities.

Patrick O'Dougherty, Truth, His Undoing: The Tiresias Complex

IRISH PSYCHOLOGY/IRISH PSYCHIATRY: THE TIRESIAS COMPLEX

Tiresias was a Theban of Greek legend who accidentally saw Athene bathing. She blinded him by splashing water in his face. Athene repented but could not restore his sight. Instead, she conferred on him the ability to understand the language of birds, and the gift of prophecy.[164]

In another version of this story, Tiresias becomes a woman and Jupiter and Juno call upon him to resolve and argument as to which of the sexes received the greatest pleasure from married life. Tiresias, testifying from experience, decided for the female. So, Juno blinded him.[165]

In a third version, Tiresias saw snakes mating and killed one of the snakes with a stick. As a result, he became a woman. The event recurred and he became a man again. Zeus and Hera consulted him about whether a man or woman received more pleasure from sexuality. Tiresias replied a woman receives nine times more pleasure than a man. Hera blinded him for this reply. Zeus gave him the gift of unerring soothsaying.[166] Like Tiresias, Patrick O'Dougherty had difficult problems with sexuality from the time of his adolescence until he was thirty-five. The death of the libido at this age brought about a personal phoenix.

In the Odyssey, Tiresias kept his prophetic gifts within the underworld where Odysseus consults him. At Thebes, he played a role in the tragedy concerning Laius and his son Oedipus. The legend is that he lived for seven generations.[167]

Truth was Tiresias' undoing. The Tiresias Complex is the psyche complex where truth is the client's undoing. For example, in Patrick O'Dougherty's commitment case, truth was his undoing. He contested Catholic Marxism and had a social justice case at Newman Center. He had to forfeit the case and accept therapy. Patrick O'Dougherty told the truth about Catholic Marxism at Newman Center to the press. This was his undoing. He has the Tiresias complex. In American history, Martin Luther King, Eugene Debs, a radical, and John F. Kennedy, like Tiresias, were undone by truth. Moreover, Patrick O'Dougherty received his certificate of graduation from Hennepin County Medical Center Day Treatment Program on November 22, 1996, the 33rd anniversary of the President John F. Kennedy assassination.

Patrick A. O'Dougherty's Case

The MMPI profile printed above presents an almost completely positive scientific analysis of my personality. God Bless Starke Hatheway, the originator of the MMPI. In contrast, the evaluation at HCMC was almost completely scientific, analytical and negative. Patrick O'Dougherty has devoted his whole life to works of charity, the mind and activism. He had an intellectual point of view and a social justice issue at Newman Center that were not acceptable. Also, he is a single person. Single people are the new majority in America, and they have few rights in America.

His Case is Before God

My case is before God. I place it before the holy temple. The psalms are my defense.

Psalm 5
O Lord, in the morning thou dost hear my voice; in the morning I prepare a sacrifice for thee, and watch.

Psalm 43
Vindicate me, O God, and defend my cause....For thou art the God in whom I take refuge;

Psalm 57
Be merciful to me, O God, be merciful to me, for in thee my soul takes refuge; in the shadow of thy wings I will take refuge, till the storms of destruction pass by.[168]

A Tribute to My Mother, Patricia Coyne

Schizophrenia is a genetic chemical disease. It is not do to poor child raising or do to a character defect. It stems from variations in dopamine levels in the brain. The writer has had problems with schizophrenia since his adolescence. It has been a descent into purgatory for him. It is his weakness, his strength and his gift. The disease runs in his family. This book and the letters that follow are a product of this disease. They are a testament to the resiliency of the writer and of his family, especially his mother, Patricia Coyne, who strengths are love, virtue, compassion and loyalty. As a tribute to mother, the writer is making her name into an acrostic of virtues and positives.

Patient
Activist
Trustworthy
Richness
Intelligent
Comeliness
Integrity
Adorable

Christlike
Orderly
Youthful

IRISH PSYCHOLOGY/IRISH PSYCHIATRY: THE TIRESIAS COMPLEX

Normal
Energetic

Prayer To Be Truly Human
"Lord Jesus Christ,
You came to earth and embraced our humanity,
thereby teaching us how to be truly human.
Help me to follow your example and so bring
out in myself all that is fully human.
Teach me to appreciate the immense good that lies
in being human,
climaxed by the gift of genuine self-giving.
Enable me to make use of all Your gifts
in accord with the purpose for which You gave them
and for the good of others.
Make me realize that only when I am genuinely human
can I be a true follower of You."[169]

The Conditions of the Solitary Bird
"The conditions of the solitary bird are five:
The first, that it flies to the highest point;
The second, that it does not suffer for company,
Not even of its own kind;
The third, that it aims its beak to the skies;
The fourth, that it does not have a definite color;
The fifth, that it sings very softly."[170]
St. John of the Cross

Like the solitary bird, Patrick O'Dougherty is a single, celibate male who flies alone. At the urban Walden, HCMC, and with his therapist, Dr. Jane Rozsnafszky, he learned, like Tiresias learned after he was blinded, the secret language of a bird or birds. His dictionary in miniature expresses that language.

"WORDS ARE DEEDS": THE NEW ABOLITIONIST MOVEMENT

The New Abolitionist Movement: An Irish Cause Therapy
Galatians 3:28 "There is neither Jew nor Greek, there is neither slave nor free, there is neither male nor female; for you are all one in Christ Jesus."

Patrick O'Dougherty echoes this passage by saying that Irish Psychology/Irish Psychiatry is about everybody. America is about everybody. He is a founder of the

IRISH PSYCHOLOGY/IRISH PSYCHIATRY: THE TIRESIAS COMPLEX

New Abolitionist Movement to make this passage and the Joycean theme, "Here Comes Everybody," a reality. This preface came out of an Easter season epiphany. It is an affirmation of himself, his family and the therapeutic process.

INTERPRETING PSYCHIATRIC HISTORY
The Revisionist School of Psychology/Psychiatry: A Critique
Dedicated To Dorothy Day, Roisin McAliskey, Anne Frank and Martin Luther King

Against the mythology generated by the agitation for personal independence, psychiatrists have created the revisionist school of psychiatric interpretation. This is the principle of 'value-free' scientific psychiatric history interpretation.[171] Professor F. Foster sees revisionism's virtue as an ability to sense half-tones, to be skeptical about imputing praise or blame, to separate present incentives from historical effects. In contrast, the basis of Catholic personalist intuitionism is freedom and represents a break from determinist, for example, Marxist or revisionist psychiatry. Revisionism is the psychiatric approach of the counter revolution in psychiatry. Catholic personalist intuitionism is the cultural revolutionary approach. Revisionist psychiatry can deal adequately with the medical models of disease. However, it cannot come to terms with the 'catastrophic' dimension of trauma, for example, the trauma of a rape victim or a victim of post-traumatic stress syndrome, or a holocaust victim. A 'value-free' scientific approach filters out the pain. Through methodologies of tacit evasion, for example, ignoring the evidence of normalization by playing down the exceptional and abnormal character of violence experienced by a socialized, colonized minority personality the revisionist approach achieves distance both emotional and moral by taking on a clinical tone and seizing on sociological euphemism and clinometric analysis which cerebralizes and thus desensitizes the trauma. All of these failings arise from the fatally-flawed adherence to the imperfect precepts of 'value-free' psychiatric history writing which denies the patient recourse to value judgements and access to the type of moral and emotional register critical to respond to human tragedy and suffering.[172]

The Mythological Foundation of Psychiatry

The documentary evidence in my case shows the complexity, ambiguity and discontinuity that characterize my personal historical odyssey. However, it is also an odyssey like that of Odysseus and Tiresias. Like these mythological figures, the writer and client has a hybrid intra and interconnected character.[173] He created a new mythological complex out of his psychiatric experience, the Tiresias Complex. Likewise, psychiatry has an intra and interpersonal mythological foundation.

Catholic Freedom

IRISH PSYCHOLOGY/IRISH PSYCHIATRY: THE TIRESIAS COMPLEX

My historical experience is a conditioned response to current personal and public issues. However, Catholic freedom is the basis of my personal odyssey. My ethnic identity, Irish Catholic, stands in continuity to the Gaelic past. This 'backward look' of my Irish personal and collective consciousness and unconsciousness turns out through introspection rooted firmly in my present consciousness, for example, in my present day psychiatric treatment.[174] The basis of my present consciousness is Catholic freedom. Consciousness has a mythological dimension, for example, my Tiresias Complex.

Personalist Psychiatric Evaluation

Psychiatric evaluation should combine a critical methodology in an analysis of the evidence with a more sensitive response to psychiatric patient content. The elements of Catholic personalist intuitionism in psychiatric interpretation are first empathy and second imagination.[175] The person is an island and the person's psychiatric records are a way of mediating through empathy and imagination the island's past and its present. The philosophy of Catholic personalist intuitionism makes the writer/client a psychiatric 'cultural revolutionary.'[176] Thus, the field of psychiatry is a revolutionary field based on the Eight Beatitudes. Catholic personalist philosophy is the integration in integrity of the person in all of his or her dimensions, for example, spiritual, intellectual, psychological, social, political, and interpersonal. For example, social integration of the person and personality in integrity forms the basis of his or her relationships. Relationships in freedom decide behavior.[177]

The Strength of Revisionism

The strength of revisionism is that it reminds us that a period in psychiatric history should be a matter for rational debate not religious or political dogmas. Out of the Catholic tradition there are many ambiguities and contradictions. As it is possible to transcend the categories of personalist and revisionist, the approaches should integrated in a dialectical synthesis.[178]

Integrative Socialism/Integrative Personalism

James Connolly is an Irish thinker who combines 'nationalism' with 'socialism.' Dr. Helga Woggon in the first biography of him in German begins with a model of 'integrative socialism' which is a special form of socialist politics developed in a situation of national and colonial independence. This idea applies to the writer's personal Irish Catholic family background which was one of personal and 'colonial' dependence on his parents.[179] His philosophy of Catholic personalist intuitionism arose through the breaking of this socialist dependence through

34

integrative personalistic synthesis. For example, the creation of the Tiresias Complex is one aspect of integrative personalism.

Ethnicity and Medical Research from Below

Ethnicity is an important element in medicine. Many ethnic groups or oppressed groups like women or gays have a 'catastrophic' dimension in their personal and medical experience. The writer did in his medical experience.[180] Medical doctors, for example, psychiatrists have given not enough attention to history or data or research from below. For example, they have inadequately researched the trauma of medical history in an ethnographic feminine dimension. In T.S. Eliot's terms, medical language is inadequate to the 'objective correlative' of trauma at the individual and aggregate level, for example, medical research on holocaust victims, the Irish white Negro and studies or writings on the slave trade.[181]

The idea of ethnicity, for example, the Irish white Negro idea as employed in the discipline of the medical, psychological and social sciences may prove to be a useful conceptual tool for reinventing psychiatry and putting it back on its covenantal foundation.[182] F. Foster depends on the stock-in-trade of the historian's mode of explanation, that is, complexity, contingency, accident and imperfection. These ideas form the basis of analyzing ethnicity and the field of psychiatry.[183] However, these methodologies should integrate at the highest level of analysis, the spiritual, covenantal level. Centering prayers, like the rosary, are one approach to this integration. Research from below integrates in prayer to the Divine above.

Scientific Self-Deception

The basic problem in medicine is the fatal result of trying with a great deal of scientific self-deception to write 'value-free' medical and psychiatric history. The dispassionate tone which concerns itself more with explaining than evoking the traumatic events, for example, the events causing post-traumatic stress syndrome has become the dominant contemporary mode of professional psychiatric or medical writing or of much psychoanalytic research in western democracies. This approach begs many important questions. It is not incompatible with an acute sense of and a sympathetic and imaginary engagement with the pain and trauma of human medical tragedy.[184] A person or a group is not 'value-free.' They have a moral dimension. A revolutionary mind set is not 'value-free.' The question the writer raises then is revisionism counter revolutionary?

The Medical Covenant

IRISH PSYCHOLOGY/IRISH PSYCHIATRY: THE TIRESIAS COMPLEX

The medical profession should engage its contents of explanation for certain critical events and processes in medical history, for example, to include the idea that psychiatry is a soul science, that the doctor like the priest is a good shepherd, that the basic relationship between a doctor and a client is a convenantal relationship which includes a promise not just between the doctor and the patient but also includes God in the promise or covenant.[185]

The medical profession's covenant should not advocate the 'normalization' of the 'tacit evasion' of the historical, for example, women's history of the reality of pain and trauma involved in depression or subjugation, for example, that is part of a commitment process. There must always be room for different perspectives, different temperaments, different emphases in the discussion and analysis of medical and psychiatric events however much these events/processes may have been marked by pain and trauma. Difference of emphasis, perspective and temperament must occur in the interests of the debate on evidence and explanation which sustains the practice of medical scholarship, and psychiatric scholarship as an academic discipline.[186]

Irish Psychology/Irish Psychiatry: Catholic Personalist Intuitionist and Revolution as Therapy

The writer is a Catholic personalist intuitionist. Personalism integrates the rights of the person in all dimensions, especially the highest dimension, the religious dimension. The personalist mind is a free "mind" building on primitive objects of intuition. These primitive objects are genetic, chemical and mathematical. Intuitionism is the activity of these primitive objects. Medical clients and doctors have personal histories and historiographies. In contrast, the medical revisionists have had a deliberate and direct assault on medical personalism which is a soul science surfacing in consciousness. Revolution as therapy is also a critical dimension of Catholic personalist and intuitionism. Capitalism is a major problem in American medicine. Many medical problems are systemic due to dysfunctions in the capitalist system. For example, the medical school at the University of Minnesota overlooks a pathology vein, the Mississippi River. It is a capitalist corporate polluters toxic waste dump. Corporate pollution causes many deaths. Thus one goal of a therapist or doctor is to put the client into a post-capitalist mind set. The whole revolutionist framework of analysis counters revisionism.

The Reaction Against Personality Consciousness Based Psychiatric History

The reaction against personality consciousness based history has developed in two phases. In the first phase, the medical pioneers since the 1930s have sought a tacit desensitization of the personalist story by looking towards the logical positivist

36

IRISH PSYCHOLOGY/IRISH PSYCHIATRY: THE TIRESIAS COMPLEX

neo-Rankean style of medical analysis. This approach is overwhelmingly test centered and favors clinical evaluation.[187]

The second phase of medical revisionism especially in psychiatry has since the late 1960s and 1970s seen the dominance of a more overtly iconoclastic approach by a new generation of doctors which in place of a positive bias towards the personalistic soul science of earlier accounts has simply substituted a negative bias as its value-base. A good example of this negative scientific, 'value-free' bias is the Diagnostic and Statistical Manual of Mental Disorders. Imperfection is crucial in any evaluation of this manual. From the Catholic personalist intuitionist perspective, paradox and irony, accident and inadvertence, inconsistency and contradiction, discontinuity and collective myth, freedom and the idealization of the past are also germane approaches to any evaluation of this manual.

Forms of Interpretative Distortion

The three main forms of 'interpretative distortion' which characterize the recent revisionist wave in medicine, especially in psychiatry are first an invincible clinical 'robust skepticism' which opts for the most skeptical possible reading possible on personal and psychiatric evidence. This results in a 'corrosive cynicism' which is brought to bear on the client to minimalize or to trivialize the importance of transcendental or spiritual operations and dynamism of the person. Second, a significant aspect of this iconoclastic assault on the soul science that is psychiatry finds itself in inverted anachronism. This is an attack on the personal myth and the person's instinct for freedom as projected backwards through his whole case study and toilet training. This reflects in a denial of the idea of a collective past. All life interrelates. This holistic conception medical revisionists dismiss as misconceived and anachronistic. In its place they emphasize complexity, ambiguity and discontinuity which characterize the client's odyssey. Third, the writer believes each client's case is a collective soul odyssey. Psychiatric consciousness is historical and the transmission of the foundations of psychiatry or bonding in the field has a mythical dimension like the Tiresias Complex which is a personal and a collective myth.[188]

The Strengths of Irish Psychology/Irish Psychiatry

Irish Psychology/Irish Psychiatry, for example, Catholic personalist intuitionism and revolution as therapy counters scientific, atheistic psychiatry and deterministic psychiatry. It is anti-determinist because it emphasizes paradox and irony, accident and inadvertence, inconsistency and contradiction, discontinuity, freedom and collective myth. Idealization of the past, for example, the psychiatric past is also important in Irish Psychology/Irish Psychiatry. Moreover, personal consciousness finds mediation through 'street' history through the 'felt experience' of authority and power, of oppression and scarcity, class, gender, and race.[189] 'Street' histories and the 'felt experience' of authority create revolutionary mind sets in clients. The key to revolution is complexity. For example, revolutions are personal, social, political, or medical. The basis of medicine is need not greed.

Patrick O'Dougherty and The Dorothy Day Movement: 'Street' Histories

Dorothy Day had an abortion. It depersonalized her. She turned to the philosophy of personalism and converted to Catholicism to spiritually, integrate in integrity this experience. This work and new medical field of Catholic personalist intuitionist interpretation adds the analytical dimension of the intuitionist school of mathematics. It grew out of my street person involvement with the Dorothy Day Centers and Dorothy Day movement in Minneapolis and St. Paul. My psychiatric case is a 'street' history.

A Note on Method: The Methodology of the Cuckoo Bird

The methodology of Irish Psychology/Irish Psychiatry is that of the cuckoo bird which lays "its eggs in the nests of other birds, which hatch and raise the young cuckoos."[190] Thus, a cross disciplinary methodology is the most desirable. My school of thought and letters analyze from the perspectives of many disciplines. Doctors raise the cuckoos!

Psychiatric Revisionism and Catholic Personalist Intuitionism and Revolution as Therapy: An Irish Psychology/Irish Psychiatry Dialectic

Catholic Integrative Personalism in Intuitive Revolution

Modern psychiatry should integrate revisionist psychiatry with Catholic personalist intuitionism and revolution as therapy. This synthesis creates Catholic integrative personalism which is intuitive and revolutionary. Catholicism has a moral intuitionist free mind basis for freedom. Medical decisions on abortion, medicide and animal implants in man should not be taken out of a moral context and placed in a 'value-free' scientific environment. Faith and reason go together in a personal dialectic. HCMC is a secular hospital. By leaving faith out of the therapy, the doctors leave out half of the dialectic. It is illegal to introduce scripture into therapy at HCMC. The patients, like Irish political prisoners, are interned. One half of the revolutionary dialectic is missing. One half of the healing process is missing. Psychiatry is a dialectic between faith and reason, for example, scientific reason. The synthesis of this dialectic resolves in God. Catholic personalist intuitionism and revolution as therapy exists in a dialectic with revisionist psychiatry. The resolution of this dialectic takes place in the relationship of a doctor, a client and God. The universe, like Martin Luther King's dream, is a dream. Roger Hallebuyck thinks "the universe is God's dream."[191] The Catholic integrative personalist process of this dream in revolution, in covenant, in virtue and in love is the synthesis of Irish Psychology/Irish Psychiatry.

IRISH PSYCHOLOGY/IRISH PSYCHIATRY: THE TIRESIAS COMPLEX

<u>ALL THESE ARE LETTERS OUT OF SCHIZOPHRENIA</u>
<u>A CASE STUDY IN "CREATIVE DIVERGENT THINKING"</u>[192]

A Product and Pouring Out of a Minnesota 'Street' History
HII

HELLENIST INTERNATIONAL INSTITUTE

A Royalist/Radical Publishing Company

"Be Ye Perfect, Therefore, As My Heavenly Father Is Perfect."
Matthew 5:48

LETTERS TO NEWMAN

9/4/94

Dr. Patrick O'Dougherty, Ph.D.
Riverside Plaza M3410
1615 South Fourth St.
Minneapolis, Minnesota 55454

Dear Friends at Newman,

Newman's <u>The Idea of a University</u>: Blasphemy and <u>Left</u>overs?

What is going on over at the Newman Center at the University of Minnesota? I went to Mass there Saturday, September 3, 1994, at the 4:30 p.m. Mass. The presiding priest was Father Rich Colgan, CSP. **A student intern, a lay person, named Chris Fellerhoff, gave the sermon or the reflection on the readings. The reading was on Christ's curing of a blind and deaf man. Chris Fellerhoff used the expression "god damn," or "god damn the fact" a number of times, maybe as many as ten or more times, to interpret this particular gospel. He ended up his sermon with the expression "god damn" for emphasis.** There were over one hundred witnesses to this sermon including Father Meinrad from St. John's. The Second Commandment, "You shall not take the name of the Lord your God in vain," and "But I say to you, Do not swear at all," was broken a number of times as part of the Mass. The Catholic Catechism is the source of this Second Commandment reference. Has the Berkeley free speech movement become part of the liturgical movement? **Are the First Amendment Rights of freedom of speech in the Constitution protected during the Mass of the Catechumens and the Faithful?** There were several Baptisms as

40

part of the Mass. Is blasphemy part of the initiation rite into the Catholic Church?
There were young children present at this Mass.

I went to St. John's University at Collegeville, and I have always been a
backer of the clergy. I go to daily Mass and Confession frequently. I am not a right
wing fanatic or left wing fanatic who wants to throw darts at the clergy. I just feel
they have gone once again too far this time at Newman.

They are behind the times at Newman. They just have **left**over leftism from
the 1960s for the young people. We are in a new era of the global market place, the
global village, the computer super highway and one world.

Another point of contention is that they have had guests speakers giving the
sermons at Newman. Former Mayor James Scheibel gave a sermon. Don't you
think Senator Dave Durenberger deserves equal time? Tom Conry, Jacqui Landry,
Jeri Cashman, OP, and Robert Hayden have given sermons there often. By Canon
Law are women allowed to give sermons? According to Canon Law, are not only
priests and deacons supposed to give sermons?

My uncle, Msgr. Loyal O'Dougherty, a former Vicar General and Director of
Vocations for the diocese of Tucson, thinks that what they are doing at Newman
Center is an "abomination." My other uncle, John O'Dougherty, a prominent
graduate from St. John's University and St. Thomas College thinks that Newman
Center should be shut down for a while. The Newman congregation are not the
Catholic College elite. They are the laughingstock of the archdiocese. The Newman
Center is financially underwritten by well over $100,000 each year by the
archdiocese. They are not getting their money's worth.

Newman Center faced censure for having an active gay group office there and
for rewriting the Eucharist prayers. Is the Eucharist still consecrated if one uses
different prayers? I think the archdiocese should deal with this blasphemous Mass
business promptly. I bet tears are coming from the eyes of the photos of Cardinal
Newman.

Dr. Patrick O'Dougherty, historian.

9/5/94

David,

IRISH PSYCHOLOGY/IRISH PSYCHIATRY: THE TIRESIAS COMPLEX

I am dedicating my book to this Irish, psychologist, Dr. Shirley Corrigan, that I had on campus as an undergraduate. She gave me the right advice about the single life, Vietnam, and dating. She is dead now. My book will reclaim part of her memory. My short story, the "Classical Father" is in the book and it reclaims the memory of a friend that I lost in Vietnam, Joseph Hart. I thought of a good title for a book on the American left, "Leftovers." My uncle John O'Dougherty suggests the title, "Left behind."

I don't want to get overly involved in personalities. It can be counter productive.

I got a book review of the new Catechism of the Catholic Church published in the Catholic Bulletin. After five hundred years of Catholicism/Protestantism in the new world, we got a few minor footnotes in this Catechism. Latin America got one reference. The French Canadians got none. The word "Liberty" was not in the index.

Dr. Patrick O'Dougherty

You are not a historian until you have lost a battle.

9/19/94

Father Steve Bossi,

Playing Trump

I sent a mail out of about twenty-five letters about the blasphemy incident that occurred in the Saturday, September 3, 1994, afternoon Mass. Archbishop Roach wrote back that he had a high opinion of the abilities of Father Steve Bossi. He suggested I have him handle the situation. Another women sent me a letter suggesting that the people involved should receive a reprimand. I sent a letter about this incident to five newspapers and one Catholic periodical. Bishop Welsh gave me a call. This should get a handle on the situation. I had a professor who sponsored a pornography show in a Christian chapel. I rapped his knuckles, believe me.

I'm for a light touch,

Patrick A. O'Dougherty, Ph.D.

Dear Friends at Newman: 12/9/94
    Revisionism and 20th Century History: The Fall II

42

IRISH PSYCHOLOGY/IRISH PSYCHIATRY: THE TIRESIAS COMPLEX

Ghandi was one of the greatest political, and religious figures of the twentieth century. This is the view from history.

Fact: So far 50 to 120 million people have died violently in Europe in the twentieth century. It was Fall/Kill [war].

The new curriculum in European schools is the metaphysics, ontology, ethics and aesthetics of murder, death, holocaust culture and population and the Fall.

A Revisionist Approach to European and Roman Civilization from the perspective of twentieth century culture: not idealism--population control. The key to population is to make it Fall.

It was 2500 years of daily bloodshed, pogroms, holocausts, drug busts, rapes, assassinations, and liquidations of dependent populations. This is daily life in modern and ancient European civilization. European Civilization is an oxymoron. One should call it violent death at any age culture: death culture or Fall culture.-----Who is one of the foremost intellectual thinkers in the twentieth century? Read the French thinker Sorel's, Reflections on Violence. He thought that every civilization has excess dependent populations that build up. This is true throughout time. One should use violence he thought to liquidate them periodically. Both the Marxists and the Fascists championed his thinking. Both groups went to his funeral(reference John Harris). Afterwards they went out and got the job done. Have they made a person with that name the head of the welfare department in France? Will his work become a best seller in America? He had a "Fallen nature."

-----The Church is right in emphasizing the spiritual life of a civilization. Materialistic civilizations, like America, pass away quickly. There is more intellectual weight in the Documents of Vatican II than in the American Constitution. This is just one of the documents of the Church. In contrast, the Anglican religion has only 39 articles. They could do a lot better.

Protestant church membership is Falling in many sects.

-----Plato thought a republic would last for only 200 years. After that time the bureaucracy would enlarge so it would not be very effective. America's 200 years are up. Virtue is the key to a Republic's survival. American history is the history of the Fall. The American is often a Fall guy.

-----There was a Civil Rights Bill right after the Civil War. Andrew Johnson vetoed it. However, most of the elements in it are in the 13th., 14th. and 15th. Amendments of the Constitution. There was very little original thinking in the 1964 Civil Rights Act or the Civil Rights Movement. The "Rights" of Man is an old idea. It should be the Civil Rights and Responsibilities Bill or Movement. Martin Luther King is the new Adam or Fall guy.

-----In most Catholic countries the people emancipated the slaves with very little violence. The King of Portugal moved to Brazil to help free the slaves. A Catholic priest argued that Lincoln was a weak leader because he did not prevent the Civil War. I think that the Civil War bordered on blasphemy. A prophet, named Abraham

IRISH PSYCHOLOGY/IRISH PSYCHIATRY: THE TIRESIAS COMPLEX

Lincoln, married a virgin named, Mary Todd, and led 600,000 people to their deaths in a white holocaust. Abraham Lincoln was then assassinated on Good Friday. He was a **Fall** guy. These are all facts. Are we supposed to think that this was not a conspiracy?

-----I love Christ, the Blessed Mother, many of the popes, saints and theologians of the Church. However, Dante put some of the Popes in hell. One wonders if after all the millions of deaths in Europe in the twentieth century how many of the Popes went to heaven. "Basilica is the Latin word for Fortress."(Reference John O'Dougherty).

-----We have had so many Southern presidents in this country and the economy has become so inflated that what they should do at the Federal Reserve is reissue Confederate currency. The debt should **Fall**.

-----The Vatican will never approve Medjugorje. You don't have a race war break out shortly after and alleged apparition of the Blessed Mother and claim that this is a true apparition. Yugoslavia had a **Fall**.

-----Go see the Frank Sinatra movie, <u>The Manchurian Candidate</u>, and then see Olive Stone's, <u>JFK</u>. Then tell me there wasn't a conspiracy. JFK was **Fall** guy. He died in the **Fall**(season).

There are more secrets in the Vatican libraries than there are in Watergate. Water**Falls**.

Vietnam/**Fall** Guys

They recently released the serfs in Europe. Soldiers, like the Americans in Vietnam, are the often serfs or **Fall** guys.

Church History/the **Fall**

The most impressive thesis in European and Church history in the twentieth century is Thomas Merton's "<u>guilty bystanders</u>" thesis on Vietnam. The clergy had relatively little dissent on any of the 125 million deaths in Europe. Were many of them guilty bystanders?

<u>The Fall</u>

I personally think the twentieth century is going to go down in history as the **Fall**, like Albert Camus', <u>The Fall</u>. The hero or anti-hero walks the other way as the woman drowns.

I think that God may give a reprieve to **Fall**en Marxists. **The key to Ecumenism, Marxism and Existentialism is mathematics. Numbers rise and Fall.**

Dear Friends at Newman: 1/9/95

Another Vietnam: William Clinton's Presidency
A Continuing Vietnam within America

44

IRISH PSYCHOLOGY/IRISH PSYCHIATRY: THE TIRESIAS COMPLEX

The North Vietnamese won the war in Vietnam, and now the American minorities are winning the civil rights insurrections When I was growing up in the 1960s, there was a moderately psychopathic president, Lyndon Johnson. He hit us with Vietnam which was a war that was never formally declared a war by the Congress. Then, he performed social experiments on us. He lost. Now President Clinton is trying similar tactics. Are the people going to accept them? Nixon was a Southerner and so was Carter. Are we are going to have a continuing Vietnam with these Southern presidents? The dollar is worth only eight cents what it was during the Johnson presidency.   Maybe they should just reissue Confederate currency. Richard Finn thinks the South is "shadowlands".

Sacrificial Love and Excellence

Christ represents many dimensions, especially sacrificial love.   Without sacrificial love there is little excellence.

Patrick A. O'Dougherty, Ph.D.

Dear Friends at Newman: 1/18/95

A New Covenant at Newman
(1) A **trillion dollar** energy conversion to hydrogen in automobiles and in all
    forms of economy: Remember the Hindenburg!
(2) A new Noble Prize category for women's contributions.
(3) A Catholic rationalization of the American experiment.
(4) A Catholic priest on the faculty of the University of Minnesota--the first.
(5) A Holy City in the New World--St. Paul.
(6) An Irish/black classic: "white Negroes": transforming the civil
    rights movement into an international insurrection. Reference: Finnbarr
    O'Doherty
(7) A Catholic sex trauma synthesis.
(8) An effort to put blood and personality back into the American races.
(9) A Catholic/Jewish synthesis on campus.
(10) A few good martyrs.
(11) Newman Center, like Oxford University, is often a center of lost causes.
    Let's change it over.
(12) A reclamation of the Catholic European **Royalist** heritage.
(13) A new field of economics: the economics of women.
(14) A Newman Center contribution to the field of the psychology of religion.
(15) An American Catholic Classic.

IRISH PSYCHOLOGY/IRISH PSYCHIATRY: THE TIRESIAS COMPLEX

Patrick A. O'Dougherty, Ph.D.

A Rough Draft

Father Steve Bossi,   1/18/95
    Patrick A. O'Dougherty and the Rights Issue at Newman Center
    As you are well aware you have denied Patrick Aquinas O'Dougherty all of his basic human rights and responsibilities as a Catholic, as an American, and as an indigenous minority at Newman Center, for example, freedom of opinion and expression, due process, privacy, and reputation To brace this rights problem and all the issues in the proposed Catholic Constitution, a parallel Constitution to the American one, I am going to look to The School of Critical Legal Studies. Reference: "Toward a Catholic Constitution," Association for the Rights of Catholics in the Church.

Critical Legal Studies

    The school of Critical Legal Studies tries to critique the <u>ideology</u> and <u>practice</u> of all aspects of the American legal system. This school adapts "a de facto ethnographic approach to legal education, the spoken and written discourse of legal professionals, and the social effects of how the system operates in practice as opposed to formal models, to which legal scholarship, closely allied with legal practice, has been prone to, but also to show how law as a process operates contrary to conventional wisdom." The basis of Plato's, <u>Republic</u>, an ethnographic dialogue, is justice. Isn't this ethnographic dialogue a basis of Newman Center, the proposed Constitution of the Catholic Church and of our American government? Critical Legal Studies tries to critique cultural hegemony, and authoritative meanings and processes. Aren't these the main problems in my complaints about Father Steve Bossi and his staff at Newman Center? It is Father Steve Bossi's problem. It is the basis of the problems at Newman Center. Reference: George E. Marcus and Michael M. J. Fischer, <u>Anthropology as Cultural Critique: AN EXPERIMENTAL MOMENT IN THE HUMAN SCIENCES</u> (Chicago: The University of Chicago Press, 1986), 154. David Noble and Patrick A. O'Dougherty are the references for the critique of capitalism.

Patrick A. O'Dougherty, Ph.D.

Dear Friends at Newman: 1/23/95
The Newman Papers
The Phoenix Experiment

    In the middle of my graduate school career I committed **"academic suicide"** and went down to Atlanta. I worked in day labor companies for a year. If you are going to be a writer, don't trust textbooks. Most of them are out of date the day that they come out on the market, and they contain half truths. You have to look to life's experiences. Atlanta will give a Northern Catholic a real education. Everyday down there was a different job and a different situation. For example, I worked at a gun convention and in different factories. I did research at Georgia Tech and the Carnegie Library in Atlanta. I applied for credit to start a business down in Atlanta. I had practically every type of situation down there: courts, colleges, evaluations, the mission, the Catholic churches, all types of racial situations, a black girlfriend, a few minor fights, and visited Martin Luther King's Church. **I told them that I had developed a complete approximate function critique of the classical equations and relativity. I got a mock Nobel Prize for this.** The Southerners played rough with Pat, but they ended up getting me a job in a Pizza restaurant. The reason I got involved with the courts in Atlanta is because my glasses broke and I couldn't do day labor very well until I got new ones. The Deep South is "Another Country." It is a shadowland--banking, politics, business deals, religion, women, everything. My Dad wanted to do, You Can't Go Home Again, with me in the South. This is a famous book by Thomas Wolfe. He is a good Southern Irish writer. Also, I did research at Widener Library at Harvard, the University of Chicago(you are lucky if you get to class alive there), Carnegie Mellon University (I like the library in the IT there), Berkeley and many other universities. I liked the University of Wisconsin at Madison, the University of Arizona and McGill University in Canada, as some of the best of the New World Universities.

Patrick A. O'Dougherty, Ph.D.

Dear Friends at Newman: 2/10/95

I'm for a Pope with a doctorate in mathematics or science.

Mixed Grades

I give the Civil Rights Movement and the Women's Movement mixed grades from results. My uncle, Father Loyola, agrees with this view also. They had a children's crusade in Europe, and here we have 50,000 plus dead children. Very few men in history have ever given the appellation of genius to a woman. I'm not exactly sure why this is true. I think there are many women geniuses.

Patrick A. O'Dougherty, Ph.D.

Dear Friends at Newman: 2/13/95

IRISH PSYCHOLOGY/IRISH PSYCHIATRY: THE TIRESIAS COMPLEX

Ayn Rand's Objectivism and the Catholic American Experience

I don't believe that the 50-60 million American Catholics should try to bail out the failed federal, state, and county programs, such as education, welfare, divorce, the debt, the military. The Catholics just do not have the funds to do this. In Ayn Rand's, <u>Atlas Shrugged</u>, the productive people, for example, the business leaders, the artists, and intellectuals go on strike. They set up a counter civilization. Ayn Rand was a Jewish emigre reacting to Soviet Marxism. Catholics should think about this. They get almost nothing on television and in the public schools. There isn't a Christian party in this country.

Faust
Faust found happiness when he got a good job as a scientist.

Moslem Heaven
I think that love making and <u>perhaps</u> warfare are a big part of the Moslem heaven. Mohammed married money and was a military victor. Winning a holy war can save your soul. The Irish don't want to convert to the Moslem religion--no alcohol.

Nearly seventy-eighty percent of the Caucasian people in this country came from descendants who migrated here after or just before the Civil War. They had nothing to do with the slavery experience here. This is true for the Germans, the Italians, the Slavic peoples, and most of the Scandinavians. They did not want slavery in most of the European countries or in most of the other countries in the world.   Can one accuse them of overt racism? It is America's Achilles' heel. Illegitimacy can destroy a people's identity.

Mpls./St. Paul
Minneapolis has a stronger economy, but, St. Paul is a friendlier, more civilized town to live in than Mpls.

Patrick A. O'Dougherty, Ph.D.

Dear Friends at Newman: 2/22/95
Realism
Lenore Scallen's son, Tom, did prison time for a money transfer. The courts disbarred her son, Steve, for withholding money in a tax case. They were all members of this congregation. Archbishop Roach went to jail for a minor offense. An Italian will do time for a minor drug offense. Senator Dave Durenberger made a minor offense of less than $5,000 and is going through an <u>inquisition</u> by the liberals. Mayor Donald Fraser and the Minneapolis City Council reneged on a contract with

48

the Society International and cost the city **$17,000,000 to $30,000,000**. They should go to prison for this. This happened during the 1930's to the City Council. There should be accountability. If Allen Welle ran up an $80,000 phone bill in a corporation, he would go to prison. Tom Linstroth thinks this.

The Military Defeat Option

My father's family did 25-30 years active in the military during WWII with five major campaigns. My brother, Mike, thinks that they need a military defeat in Appalachia. My nephew, Sean, an eight year old, thinks Los Angeles needs an invasion. Nicole Brown Simpson is a Catholic. O.J. Simpson pulled every trick in the book on her. My uncle, John, thinks that the South should be given another military defeat--two to three million dead. You can't change human behavior very easily. "If you are having trouble with a town, blow it up."(Ref. Cliff Strophe) This happened in Spain and Italy and many European towns, for example. Patrick A. O'Dougherty, Ph.D.

Dear Friends at Newman: 2/23/95

On Former Mayor Donald Fraser Coming to Newman
Minneapolis/A White Elephant

Hubert Humphrey and Arthur Naftalin were the good mayors in Minneapolis. I don't have an axe to grind. However, Donald Fraser has been a disaster. He has championed many ACLU anti-Catholic causes in Minneapolis. Donald Fraser makes the ACLU look right wing. LSGI is a French probably Catholic name. He ran it out of town. Moreover, every neighborhood in Minneapolis is a disaster except those along the river road and those near 50th and France. Much of downtown Minneapolis is an economic **white elephant**. For example, the Target Center is a "**white elephant**." Also, there is a 9 1/2% sales tax on **food and drink** in downtown Minneapolis. Minneapolis is a beautiful city. However, Minneapolis became a spiritual, intellectual, artistic, political, economic and social **white elephant** under his administration.

Shakespeare and the Lawyers, especially Fraser

Senator Dave Durenberger and Tom and Steve Scallen (liberals), lawyers from Newman Center, made minor mistakes and got put through an inquisition in the courts and in the press. Also, several members of the city council during the 1930's went to prison for minor offenses. Donald Fraser and the city council, in contrast, wrecked this town and cost this town millions and got away with it. The reason lawyers are afraid to challenge Donald Fraser is because his father was the head of the law school. What about the law school? I used to have a high opinion of law and lawyers. However, there are more lawyers in Minnesota (around 16,000) than in the country of Japan. Today we are witnessing the "corruption and destruction" of the American legal system. For example, the law school at the University of Minnesota has become "an intellectual house of prostitution." I would turn down a tenured position at the law school at the University of Minnesota.

IRISH PSYCHOLOGY/IRISH PSYCHIATRY: THE TIRESIAS COMPLEX

To the next American Revolution: "The first thing we do, let's kill all the lawyers." William Shakespeare, Henry VI, part II, act IV, scene ii, lines 83-84. Dick the butcher is speaking. The lawyers in this country need to come to terms with Shakespeare. The law school and the legal profession are another Vietnam. The history of the American legal profession has become the history of "usurpation."

Patrick A. O'Dougherty, Ph.D., Intellectual Historian.

Dear Friends at Newman: 2/27/95
An Intellectual Massacre at Newman
Mark Danner gave a talk on his book, <u>Massacre at El Mozote</u> at Newman. It was about a left wing military loss in El Salvador. He almost completely appealed to emotion rather than to reason. He should have at least tried to look at the legal, political, philosophical, and Catholic questions involved. Didn't he take high school philosophy? Intellectually, he shot himself in the foot, and I cried. It was another example of liberal **death culture** as opposed to Christian culture. The liberals have control of more than eighty percent of the mass media. Death and scandals are their American diet. Reagan got diplomatic recognition of the Vatican for America for the first time in about 100 years. Many of the secularists opposed it. Did Mark Danner ever bring up directly the Catholic question?

Charlie McCarty: A Good Mayor in St. Paul?

Charlie McCarty was a police officer in the late 1960's and 1970's who became mayor of St. Paul. He made many initiatives in his tenure. However, his main effort was to build up the television and movie industry in St. Paul and in this state. He made a couple of movies and had a television show here. Thanks to Charlie, we got a movie industry going here. Did lawyer mayors Larry Cohen or James Scheibel do this? Charlie McCarty took some big risks.

Aquinas O'Dougherty thought secular education had **"nothing to stand up and be counted for."**

Catholics and Catholics schools have ideas to "stand up and be counted for." In contrast, most secularists and many Protestants and Jews have **many ideas**, but they are <u>often</u> all about equal. I went to both private and public schools, and I can testify to this.
Patrick A. O'Dougherty, Ph.D.

50

IRISH PSYCHOLOGY/IRISH PSYCHIATRY: THE TIRESIAS COMPLEX

MOVEMENT FOR THE ABOLITION OF FEMALE CIRCUMCISION
MAFC
Founded by Patrick A. O'Dougherty, Ph.D.
4/26/95
Newman Center University of Minnesota

This movement is for the nobility of all women, especially African and Arabian women where female circumcision is a common problem. Frank Wright, M.D., is the reference for Arabian women and their female circumcision plight.

THE CATHOLIC SCIENTIST ACTIVIST MOVEMENT
CSAM

Founded June 9, 1995

by

Patrick A. O'Dougherty, Ph.D.

University of Minnesota Newman Center

Dear Friends at Newman: 8/2/95
Newman Activities: Math as "Activity"

There is a continental mathematician named, Jan Egbertus Luitzen Brouwer, who invented the intuitionist school of mathematics. It is the only school of mathematics that is still standing. The theme is that mathematics is an "activity"-- "free ('mind') building mathematical objects beginning from primitive intuition." In contrast, the Platonist think math is ideals. I relate the mathematical activity theme to female intuition, Catholic personalism, for example, Dorothy Day and Pope John Paul II, the Paulists, Liberation Theology, the Irish experience, the holocaust, evolution, American women, Simone Weil, Aquinas, Thomas Merton and Newman. Brother Gregory Conant relates the activity theme to creation in <u>Genesis</u> 26-27. I counter Marxism with personalism and intuitionism. A reference to intuitionism is in Michael Dummett's, <u>Frege: Philosophy of Mathematics</u> Cambridge: Harvard University Press, 1971. Michael Dummett is an English Catholic convert who worked with minorities in England and taught Father Hilary Freeman who lives at Holy Rosary Parish at Oxford.

Patrick A. O'Dougherty, Ph.D. personalist, intuitionist

IRISH PSYCHOLOGY/IRISH PSYCHIATRY: THE TIRESIAS COMPLEX

Editorial Department
Attention Cindy Paslawski
The Wanderer
201 Ohio Street
St. Paul, Minnesota 55107

Staff:

8/18/95
Catholic Marxism at Newman

Father Steve Bossi is a new priest at Newman Center. He came here from Clemson University. He is the first Italian priest to head Newman Center. He is a Catholic Marxist from what I can tell. So are many of the Catholics in Italy. The Archdiocese of St. Paul underwrites Newman Center $196,000 a year. Do they underwrite the Wanderer that much? St. Joan of Arc is a popular parish and so is St. Agnes. Newman Center is not. It is practically empty six days a week. Father Rich Colgan, the other priest at Newman Center, does not welcome a brace from The Wanderer. Tom Conry is the liturgist at Newman Center. He offers Brechtian liturgies. Brecht was an East German Marxist dramatist. The liturgies are like Greek dramas. See the enclosure on "Looking at Liturgy: Aristotelian Vs. Brechtian Lenses." Tom Conry's liturgies represent the death of liturgy. I asked Father Steve Bossi or Father Rich Colgan to invite The Wanderer to Newman Center. They wouldn't hear of it. He drew a line with me--no concessions, no recognition, no dissent on everything including race. Is this John Henry Cardinal Newman's, Idea of a University? I like John Cardinal Krol as a Catholic thinker. I think the Church should canonize Cardinal Dennis Dougherty.

Tom Conry: Newman Center University of Minnesota Liturgist
"Looking at Liturgy: Aristotelian VS. Brechtian Lenses
(With Minor Editing)

	Aristotelian	**Brechtian**
Opening Rite	sets mood	states ground rules
Service of the Word	dazzles w. story	relates story (catechesis)
Preparation of the Table	maintains mood	reflective, of critical pause
Service of Table	dazzles w. action; catharsis	decision by the assembly; inflames

IRISH PSYCHOLOGY/IRISH PSYCHIATRY: THE TIRESIAS COMPLEX

Closing Rite	denouement; had experience	sum up argument experience is to come
Presider	"becomes Christ"; talent dependent	"does Christ"; ritual dependent
Lector	"becomes" story now	relates ancient text
Homilist	individual eloquence	proclamation of good news
Cantor	romanticizes psalm	makes clear argument
Choir	creates mood	asks common question

ARISTOTELIAN	BRECHTIAN
feelings are manufactured	feelings are a by-product
reverence	hospitality
individual salvation	solidarity
appeal to another "world"or reality	appeal to this world, this reality
suspension of disbelief	a realistic skepticism; "fides quaerens intellectum"
relates to "inside"	relates to "outside"
concern for person	concern for assembly
music dishes up	music communicates
music heightens text	music sets forth text
music paints situation	music gives attitude
provides an experience	makes an argument

IRISH PSYCHOLOGY/IRISH PSYCHIATRY: THE TIRESIAS COMPLEX

provides sensations	makes a decision
concern w. beauty	concern w. truth
preserves, affirms culture	challenges culture
assembly taken for granted	assembly object of song, story, and action
assembly "involved"	assembly faces situation
assembly is "colony"' a "client class"	assembly is controlling body
"flow"	discrete liturgical elements
assembly ought to imitate presider	assembly is font of all other ministries (including that of presider)
implicates assembly in work of the presider, but; wears down capacity for action	makes assembly "observers" sometimes arouses capacity for action
affective path to "whole person"	rational path to "whole person"
instinctive feelings manipulated, stimulated, enjoyed	feelings are examined
feels good	feels like doing good
justifies words and actions by (past) causes	justifies words and actions by (future) purposes"

Patrick A. O'Dougherty, Ph.D.

Dear Friends at Newman: 11/26/95

Anti-Catholicism in America: The Main Stream

   Contradiction of the Catholic and Christian synthesis is the basic premise of secular education (an oxymoron) in America.

IRISH PSYCHOLOGY/IRISH PSYCHIATRY: THE TIRESIAS COMPLEX

The real bastion or "last bastion" of anti-Catholicism in America is the secular school system which is basically an atheistic/agnostic school system. This school system feeds right into the media which basically locks out Catholic culture in America. The central premise of the University of Minnesota is to counter the Catholic and spiritual premises. The Catholics pay for this school system and what is on the media. The most anti-Catholic country in Western Civilization when it comes to Catholic culture which bases itself on the Incarnation is America. Catholics have a much better lot in Canada, Ireland, England, France, and Germany than in America. The liberals are a major source of anti-Catholicism in America. There are others. There is almost no evidence to the contrary. There are only fifty-sixty million Catholics in America. How can the social programs of this small a number of people possible compete with the spending power of the federal government? They can't possibly. The liberals in this country want to triple the burden on the Catholics and make the Catholics bailout their failed government programs. Many of the Protestant sects in this country do not have private schools. This is one of the main reasons they do not want to give the Catholics any concessions in the schools. Brother Gregory Conant contends "not one Protestant denomination, as a group, made any formal concessions to the Catholics." The groups most noted for being anti-Catholic are the Unitarians, the Presbyterians, and the Baptists. Many Jews and many people in the ACLU do not favor giving the Catholics and the Christians any concessions in the secular school system. There are some exceptions to these observations but they are rare. The Christian groups most willing to work with the Catholics are the Episcopalians, the Lutherans and some of the Jews. The Catholics, in America are noted for "paying taxes all out of proportion to their numbers, bearing the military duty all out of proportion to their numbers" and much of America's trade is done with Catholic countries. There really isn't much academic freedom in America about Catholicism.
Patrick A. O'Dougherty, Ph.D.

Dear Friends at Newman: 2/20/96
Death Culture
The Democratic Party in America is the "war party": Jefferson, Robert E. Lee, Wilson, Roosevelt, Kennedy, and Johnson. The Republicans want to stay home and make money, open trade routes or work on diplomacy. Promotions come very slowly in the military or government during peacetime. The culture of the American liberals is "death culture." This is a big theme in literature. Postcolonialism is another big theme in literature. Many American minorities, especially Catholics, are colonized here. After the O.J Simpson trial, America has become "Balkanized."

Changing Demographics
Demographics of the U.S.A. 1990 Census
From Doug Henwood: The U.S.A. Atlas

IRISH PSYCHOLOGY/IRISH PSYCHIATRY: THE TIRESIAS COMPLEX

German 23.3%
Irish 15.6%
English 13.1%
Afro-American 9.6%
Italian 5.9%

However, if you count all the Latins together, for example there are nearly 17 million Mexicans, I think they have a greater population than the English. There are more practicing Jews in this country than Episcopalians. The WASPS used to be 85% of the population of this country now they are 13.1%

Homosexuality: Pyramiding

The leading cities in the homosexual movement in America are San Francisco, Minneapolis/St. Paul, New York and Washington, D.C.
Minneapolis/St. Paul are the Cities of the Plains: Sodom and Gomorrah. Homosexuals pyramid,for example, in the military. They have the DFL in Minneapolis locked up.

Patrick A. O'Dougherty, Ph.D.

Dear Friends at Newman: 4/25/96
Quebec Independence
A Separate Military Option For Catholics
The End of Cowboy Roundups of Mexicans and Latins

What rules the world is force. When Catholics have a separate military option in the New World, people will leave them alone. They will then express their culture. This is what French identity in Canada will provide them. Until that time, Catholics will just be "white Negroes" here. Back in the early 1950s one million Mexicans and Latins were rounded up an shipped back to Mexico. It was a cowboy roundup. This is going to continue until the Catholics have a separate military identity in the New World. The Southerners are going to force a Catholic identity in America.

Regional Governments and City States

America should have regional governments and city states with different philosophies. For example, Canada has around twelve provinces or regional governments. There are economies due to scale. This will happen eventually in America. Minnesota has alone eighty some counties while Massachusetts has around fifteen. There should be around eight to twelve economic regions in America. This would make much more sense. The Germans tried 300 principalities for centuries and this did not work out for them at all.

56

IRISH PSYCHOLOGY/IRISH PSYCHIATRY: THE TIRESIAS COMPLEX

The Metric System: "The French Connection"
France introduced the metric system. England has finally accepted it. So has everybody else. A lot of our trade to the Third World just goes to replacing nonmetric parts for our products in those countries. American conversion to the metric system will enhance trade. Why should the Japanese buy nonmetric parts for American automobiles sold in Japan? It is ridiculous.

Patrick A. O'Dougherty, Ph.D., Part French Canadian.
Walter Disney was an Irish Canadian.

Father Steve Bossi,   4/26/96
No Middlemen
No regrets. I took this as a compliment. I didn't know you had interest in a male like me. I had a mail out on this situation and filed the necessary reports. The copy is enclosed. Let's set up a meeting on Monday and go over this little tiff. Let's see what we can work out. If you have forgotten the number of the Campus Police, I looked it up. It 624-3550. If you feel more comfortable with the security at the meeting, call them. You have my written permission to call them anytime. However, I'm for no middlemen. Are you? You can't make money in business deals if you involve a lot of middlemen.
The Limitations of Personalities
The deal is you have a strong personality. It makes for a good leader. However, it isn't a warm personality. The problem is you are not good on the one to one. Intimacy is a big problem for many people. I don't know Father Forliti very well. However, he doesn't appear to have this problem. The idea with human weakness or the particularities of personalities is to discount them. Try it. Discount your own first. So, you don't lose face, I am going to come around and see you on 1:30 P.M. I don't want any big war over this. Look at this from my point of view. If you have a problem with this, call me up at 339-1748. Do you have any remedies for this? I should get something out of it--a win-win situation.
Steve Bossi Name is On This Report Nobody Else's
If you try some stunt over at Newman, I will seek <u>criminal</u> as well as civil penalties against you. I could always plug in some real big dollar signs. Then, I could float this to different newspapers. The chancery didn't pick a fight with me. I want the 1:30 p.m. time. So, I have time to file the lawsuit before the government offices close. In case there is some stunt. I want the daylight time. I wasn't read my rights. The trespassing notice wasn't correctly executed. By in large in life you reap what you sow. If you want to sow discord, no rights, no positive recognition, with me then you will reap that.
Patrick A. O'Dougherty, Ph.D. I will keep this strictly between you and me until we have our conference. Also, I have a right to privacy. See enclosure.

57

IRISH PSYCHOLOGY/IRISH PSYCHIATRY: THE TIRESIAS COMPLEX

Father Steve Bossi,   4/28/96

The Male Resonance Problem
I have a family with many counselors in it. Maybe your family doesn't. You cannot able to handle sparks, male resonance, understandings, and resolution? Would you give somebody a major appointment in most areas of life that had that problem? The person might just say a few friendly words to you and deceive you, or you might end dead in combat out of it. Would you give somebody and "A" grade in graduate school you didn't have male resonance with? You are a priest confessor and you can't handle male resolution. I would think about this problem. Why do you have this problem? This is the problem here, right?

Patrick A. O'Dougherty, Ph.D.

Father Steve Bossi,   5/1/96
Friday
I am going to come around to your rectory about 11:00 A.M. on Friday. Let's see if we can resolve this. If you don't do this, we will have a friendly court case? I don't know much about women. I do know a lot about men. If a male is too friendly with me or too formal with me, I don't do it. I feel you are too formal with me. I know little about neurology. I would be fearful if a car accident like the one you had happened to me.

Patrick A. O'Dougherty, Ph.D.

Father Steve Bossi: 5/1/96 Evening
Five Years of Subtle Negativism With Patrick A. O'Dougherty, Ph.D.
I made an appointment with a Catholic woman psychologist today to ask her how to resolve a situation of subtle uninvited negativism with me at Newman Center. I told her several weeks ago that I was going to try a lawsuit. I gave her a copy of the letter you gave me. I have never made any personal or social threats over there. She told me to make the these points. I just mentioned a lawsuit to the City Attorney. I do not want to pursue a lawsuit. I also want my privacy protected. I told her that I contested the Newman budget last year at the chancery.   Father Steve Bossi contacted Dr. Joseph Gendron about this. She told me that he had no right to do this and that Dr. Gendron released no information. People cannot legally call up a priest, or a psychiatrist or a psychologist and ask for personal information about a client. She told me not to release her name to Father Steve Bossi. She told

me to tell him that he is not welcome with my doctors. This is perfectly legitimate. She also thinks that it is not easy to resolve the intellectual issues at Newman Center. She thinks that you should agree to lift the trespassing ban which you asked for to avoid this litigation which I do not want. I personally want no more negativism from either side. I had a perfect right to publish an article in the Minnesota Daily about using the expression "God Damn It" ten to fifteen times in a sermon. I also had a perfect right to float an article to several newspapers about Brechtian liturgies at Newman Center. He was an East German Marxist Dramatist. I want this situation resolved in a situation of mutual respect. This psychologist thinks intellectual and personality issues are often the most difficult to resolve. She recommends different channels for my provocative ideas. She also told me my doctors are not Father Steve Bossi's concern. I also told her that Father Steve Bossi is a military man. He invited a lot of negative feedback from an intellectual historian. She agreed with that. I also told her that Father Steve Bossi is a strong leader. However, he finds it difficult to work on the one to one. For example, he has a problem with masculine resolution: sparks and understandings. She said this is probably true, however, don't rub it in. She suggested that I go to Father Steve Bossi and bring up the idea of reconciliation and resolve it that way. I also told her that I don't give out many compliments to people. She told me that I should give him some positive feedback. Has he ever given me any? She also thought I should continue to work on my library and to pursue other intellectual channels. I am not going to press the pocket nerve any more. However, did I ever get any money from Newman Center?

Patrick A. O'Dougherty, Ph.D.
cc. Dr. Joseph Gendron and Father Mike Tegeder

Father Richard: 8/13/96
Father Steve Bossi's Last Hurrah at Newman Center
     I saw Michele Gersich in Dinkytown today. She told me that both you and Father Steve Bossi are leaving and that Sunday is Father Steve Bossi's last Sunday at Newman Center. I have some feedback for you on Father Steve Bossi's last hurrah. Father Steve Bossi is a good priest. He is a good leader. He is an excellent administrator. He has strong suits in social justice and race relations. He gave credible sermons. I liked his newsletter. His liturgies were crisp and clean. He has a high level of morality. He ran a tight ship at Newman Center and a tight ship is a happy ship. He chose people of high caliber to work with him. The congregation received him well. The problem that I had with him is that he isn't strong in counseling, confession and psychological mindedness. He was always subtly negative with me and overplayed the race card. Also, it was the military man versus the writer activist. I don't consider it is a good idea to rub a writer the wrong way like he chose to do. Both of my parents were counselors and I have devoted my life

IRISH PSYCHOLOGY/IRISH PSYCHIATRY: THE TIRESIAS COMPLEX

to writing, activism, analysis, law, business and the church. It just appeared to me to be wise to opt for a professional evaluation and third party intervention with the administration at Newman Center. This approach is more objective. It gets the job done. It is fair. I give Father Steve Bossi an "A/B" borderline for being a priest. I hope that he likes Portland where he can be closer to his friends and family. Why don't you give him some good counseling advice from me? Tell him to lighten up a little bit. Honey goes farther than negativism. You are also a good priest.
No hard feelings,

Patrick A. O'Dougherty, Ph.D.

Confidential

Fintan Moore, Campus Minister
Newman Center
1701 University Avenue S.E.
Minneapolis, Minnesota 55414-2076

Fintan,
I saw my psychologist, Dr. Jane Rozsnafszky, last week. She told me it was a good sign that I wanted to go back to Newman Center. She suggested that I go over to Mass on Sunday and greet the people and keep a low profile which I did. I told her that you had a talk with me and wanted to know the upshot of the court hearings. She suggested that I send you a copy of the court hearing and to tell you that I am following it carefully. However, she notes that it does not say that I cannot go back to Newman Center. It just says that I have to avoid Father Steve Bossi who left. I feel that this departure has cleared the air. The problem that I had with Father Steve Bossi was that I could not break the ice with him and have male resonance and sparks. I have taught in graduate school, and if you have a student like that where you don't know where you stand with him or her you don't give them an "A". You personally don't do that student. I have been a counselor and if I don't know where I stand with a person. I don't do him or her. If there is a question mark, I don't do it. What I want to do at Newman Center is to resume going to Mass there. I have been a member on and off since the 1960s. I am an alumni of the University of Minnesota, and I taught there. It is my home. I want to recapture the spirit of Christianity at Newman Center, focus on my library, and my books and develop some new directions. I want to emphasize the positives at Newman Center and move on in my life. I also don't want to get overly involved in personalities. I wish the new priests well. I have developed a very large Catholic synthesis which I think sets a fine precedent at Newman Center. I have several academic invitations which I want

60

to focus on. I personally wish you and the staff well. You took the initiative with me, and I welcome your initiative. I don't hold any hard feelings against Father Steve Bossi. I gave him a good grade. He told me that he liked me. I want the power struggle dropped. It is counter productive for my endeavors. At the court hearing, they told me that I should get to continue to keep doing the IASA library. Dr. Rozsnafszky thinks that I'm strong in Catholic faith and in Catholic charity, but I need work in humility. She thinks I need a spiritual advisor. I have a normal MMPI and did quite well on all the tests given me. The reason that I sought counseling as an undergraduate at the University of Minnesota was not due to a whole lot of family problems or personality pathology, but to keep from falling though the cracks in life and at the university. See you Sunday.

In the spirit of reconciliation and renewal,

Patrick A. O'Dougherty, Ph.D. prodigal son, 339-1748

I want this kept confidential. I have a right to privacy.

Fintan Moore, Campus Minister
Newman Center
1701 University Avenue S.E.
Minneapolis, Minnesota 55414-2076

Fintan,
"The Wretched of the Earth"
My psychiatrist mentor is Franz Fanon who wrote The Wretched of the Earth. I'm a Catholic intellectual who ended up having Viktor Frankl experiences in life. I want to keep my situation at Newman Center open ended. Dr. Jane Rozsnafszky's phone number is 377-2302. She is a Catholic. Call her if there is a problem. She is the one who gave me the advice on how to handle the situation with Father Steve Bossi. Do you think Father Steve Bossi treated me in a Christian manner? I have many accomplishments in life. I published eight books in intellectual history, invented a new school of physics, built a research library on campus, and created a psychiatric archive. All that I claim is that I'm a sinner. I don't believe that I'm not going to get credit for what I've done. You reap what you sow in life. The chickens will come home to roost. I paid a terrible price in life. I'm an alumni. The doctors have always been for me. They just think that I should tone it down a little bit. I put in seven years of daily labor in at Newman Center gratis and got no credit for it. The key to peace is justice. I just called a black civil rights attorney about having no rights at Newman Center, and he told me to call the City Attorney about it. The City

61

IRISH PSYCHOLOGY/IRISH PSYCHIATRY: THE TIRESIAS COMPLEX

Attorney told to file MP police reports on it so I did that. I guess Father Steve Bossi didn't like that, but that is what I instructed to do. I hope this resolves it. I don't wish anybody <u>any harm or ill will over this</u>.

Like Moses I'm "A Stranger in a Strange Land,"

Patrick A. O'Dougherty, Ph.D.

Father Paul Rospond, CSP
Father Charles Martin, CSP
Fintan Moore and Newman Center Staff
                    12/10/96
    "Knock the Dust Off Your Shoes!" Joyce Forsgren, coordinator

        The case against Patrick A. O'Dougherty, Ph.D., was dismissed on November 13, 1996. Janis Vape, psychologist declared "all government proceedings have been terminated," and "your services have ended according to your social services plan." Patrick O'Dougherty has been cured and rehabilitated. He has experienced a miracle. His university professors told the counselors that he was "brilliant." Patrick O'Dougherty has decided to move on in life. He wishes to retain the right to attend a funeral at Newman Center should any of his friends die there. He wishes to keep the Newman Center situation open. He is going to edit out the negatives in his letters and publish a first rate book out of the experience. Patrick O'Dougherty thinks the Catholic Civil Rights Case which he had with Father Steve Bossi was one of the best things that ever happened at Newman Center.
God Bless,

Patrick A. O'Dougherty, Ph.D.
See enclosure.

cc. Dr. Jane Rozsnafszky

Father Steve Bossi: 10/4/96
                    At Loggerheads
        I called the City Attorney and asked him what do you do about a loggerheads situation with a priest in campus ministry at Newman Center on campus. He recommended filing a complaint with Minneapolis Police Serv. I did this on 10/2/96.

IRISH PSYCHOLOGY/IRISH PSYCHIATRY: THE TIRESIAS COMPLEX

A minority woman called me up that evening and took down the complaint. The case number is MP96092631. I raised the different issues about freedom of information, rights of a Catholic in decision making and dissent, rights of a single person, minority issues, personality clashes, fair play, and campus criminal ministry. There are several different courts you can file a case in, for example, criminal, conciliation, and civil court. I checked them all out. The clerks recommended civil court. The filing fee is $132.00. Bill told me you transfer this year. A summons and complaint are filed. A disinterested third party serves the complaint. The party complained against, Father Steve Bossi, has sixty days to reply. The law library has lumina. The case goes on a court docket. It comes up after a short while. I like rough stuff. However, I want to get something out of this. Psychologists are usually for win-win situations. Oftentimes in the courts they do stealing from Peter to pay Paul. What do I get out of it? I could file this law suit on Cinco de Maio.

The Chickens Will Come Home to Roost,

Patrick A. O'Dougherty, Ph.D.

Nickel/Dime Observations
My dad thought the history of the Roman Empire is simply "the armies never stopped marching out of Rome." Also, there was a pogrom somewhere in Europe every twenty-five years. Donald Webb thinks European history is largely "the rise and fall of the Roman Empire." For example, the statement of the Irish people is mainly "Christ and Caesar."
-----Ireland is Europe's "poor old woman"--not a lot of money but a lot of culture and wisdom.
-----The Southerners, many of them, want to "get even" with the Northerners for the Civil War. They have.
-----Franklin Roosevelt had a blockbuster bomb dropped on every small town in Germany. He wanted to give every German a taste of the war. Tacitus thought the Germans are a very warlike people. They want to fight you for it. The Civil War was just a shaving cut compared to Europe's war deaths in the twentieth century.
-----It looks like the bottom fell out of the Russian race.
-----The Catholics have an aristocratic heritage in the world: music, architecture, art, science and religion. Americans think we are relics. They don't have a count Chopin in the South.

-----My dad served 11 years in World War II. He wanted nothing to do with the Europeans. He was for the common man and interested in third world peoples.

IRISH PSYCHOLOGY/IRISH PSYCHIATRY: THE TIRESIAS COMPLEX

-----In many ways America has become an absurdity. It is a business and <u>pedestrian</u> civilization.

-----I like Tom Conry's guitar liturgies at Newman. The music is easy and catchy. However, that is not what Beethoven was all about. Practically "everything is tough with Beethoven."
-----The Irish are quite individualistic compared to many races.
-----I don't want to step on toes, however, I love dissenting essays.
Patrick

Father Paul Rospond, CSP
Father Charles Martin, CSP
Fintan Moore and Newman Center Staff
12/10/96
"Knock the Dust Off Your Shoes!" Joyce Forsgren, coordinator

The case against Patrick A. O'Dougherty, Ph.D., was dismissed on November 13, 1996. Janis Vape, psychologist declared "all government proceedings have been terminated," and "your services have ended according to your social services plan." Patrick O'Dougherty has been cured and rehabilitated. He has experienced a miracle. His university professors told the counselors that he was "brilliant." Patrick O'Dougherty has decided to move on in life. He wishes to retain the right to attend a funeral at Newman Center should any of his friends die there. He wishes to keep the Newman Center situation open. He is going to edit out the negatives in his letters and publish a first rate book out of the experience. Patrick O'Dougherty thinks the Catholic Civil Rights Case which he had with Father Steve Bossi was one of the best events that ever happened at Newman Center.
God Bless,

Patrick A. O'Dougherty, Ph.D.
See enclosure.

cc. Dr. Jane Rozsnafszky
Bishop Lawrence Welsh: 1/8/97
The Newman Books
I wrote a positive work out of my experience at doing library building at the International Alliance for Sustainable Agriculture at Newman Center. I called it <u>St. Patrick, The Green Revolution and the Hydrogen Conversion Project</u>. I am the main founder of the Green Revolution at the University of Minnesota. I am going to see that you get a copy of it. I would like you to hang on to this copy. However, as you

are well aware, I had a real lot of tough experiences there. I am writing a second book out of these experiences. Basically, what I am going to say is that I experienced a recapitulation of the Thoreau experience in civil disobedience and ended up with an urban Walden psychiatric experience. I am going to say that I was "scapegoated" for five years. This is true. This may sound a little tough, but it happens to millions of people in this country and around the world everyday, especially to women and children. I'm sorry that this also happened at Newman Center. I got a book called Green Rage: Radical Environmentalism and the Unmaking of Civilization by Christopher Manes. I am going to take the terms in it and translate them into a Green Psychological/Psychiatric Dictionary and place them into this experience. The letters that I sent to the chancery are the seeds of this experience.

God Bless,

Patrick A. O'Dougherty, Ph.D.

Bishop Lawrence Welsh: 1/15/97
"Blessed Are The Poor In Spirit"
A Catholic Thoreau Experiment in Minnesota
While I was a student at St. John's, I got involved in counseling and in Student's for a Democratic Society radicalism. I got involved in the anti-Vietnam war movement up at St. John's. I majored in history. However, they had compulsory ROTC when I was a freshman at St. John's in 1964-65. My family had twenty-five to thirty years active in WWII with five major campaigns. I was an army brat at Fort Custer, Michigan. I read the radical material on the Vietnam War and I did not think that the Americans were going to win it. I also wanted to try something different from the military. I sought counseling about this and about some personal issues and transferred to the University of Minnesota during my sophomore year. I had great luck at the university and sought counseling there also. I saw an Irish psychologist, Dr. Shirley Corrigan, who had been a medical nurse in the navy and had been stationed in Corpus Christi, Texas. When the draft situation came up, I indicated on the forms that I was seeing her. I told her that I didn't feel comfortable with the military, and I didn't want to fight in a colonial war because the Irish were a colonized people. Also, Bobby Kennedy had just been slain. She told me to have the doctors at the Draft Board contact her. She just told them that I was not well suited for the military and that I was better suited for academic pursuits. The Draft Board accepted her recommendations and that ended it. The dissidents prevailed. I later descended into psychological and psychiatric purgatory, but I have made a lot of significant contributions in life. I want to pull my story together in a powerful Catholic dimension. The counselors that I have now want to encourage me.

IRISH PSYCHOLOGY/IRISH PSYCHIATRY: THE TIRESIAS COMPLEX

However, the point Dr. Werner makes is what if somebody does not want to do this with you. They have a right not to. Do you want to do this with me? You are your own man. Priests counsel soldiers. Why wouldn't they emphasize counseling psyche consumers? If you don't want to do this, I want to find another Catholic priest to do it. I want a first-rate Catholic priest to do work with. I can move on. In fairness to me, what do you personally want to do with my case? There has to be something for me in the relationship as well. God Bless, Patrick A. O'Dougherty, Ph.D.

LETTERS TO FAMILY

Teeny Reeny, the Genie, 2/27/94

Thanks for the whistle. I've been using it to whistle at Major, Mary Ann and Steve's dog. I talked to Mame yesterday. She told me that she liked Rio. I sent Megan and Sean some affirmation stickers. An example is the following: "I like your choice of your adult vocation." Stefan wrote a fifteen page book about some characters named Snoopy and Woodstock. Ross sold some writings. I heard that Jorge got a job in a used bookstore. Do Brazilians use much birth control?

I am reading a book on process-psychology called, Riding the Horse Backwards. It organizes different techniques in the various psychology schools like Jungian, Freudian and Gestalt into various therapeutic processes. Is E-mail expensive? As a people, do you think that the Brazilians are quite "carnal?" Is there a Brazilian Kinsey Report? I gave mother a pearl pendant for her birthday. Why don't animals live lives as celibates?

Privacy

I would like my right to privacy respected about my doctors and my past. By the time when Stefan, Sean and Megan are teenagers, I will be at retirement age, 55. Why tell them about all the psyche problems? I want recognition for my positive contributions. I want respect when I'm older.

Rose is in a nursing home. Steve lost his job when Riverplace closed, but he got a new gig at the Steak and Ale in Bloomington. Are you going to reconcile with Catholicism? When are you coming home from Brazil? My book on Shakespeare is almost half done. What is your favorite Brazilian food? I heard from Mame that you have a three bedroom apartment. Is it beautiful? Are you still doing interviews?

I went to seminar on Philip Greven's book, Spare the Child. It is about a nonviolent approach to child rearing. One of the themes is that violence can create a reaction formation personality like the authoritarian personality in the adult. Also, we can not use violence on prisoners, but we can use violence on children. Did you watch the Olympics? I was rooting for Nancy Kerrigan. However, I don't work out with weights or skates. Instead, I work out with ideas. Mike Franey left <u>Formula</u> and is going to try to start a new company. He is taking some time off. I have several friends who are retired already. I hope that I've raised enough questions with you.

Love,

Dr. Patrick

Tiny Rene or Maureen or Mo or Renee,   5/18/94

        I hope to start a Catholic Studies Department at the University of Minnesota. It would be the first in the nation. The secular university is the last bastion of anti-Catholicism. It would be a significant contribution. I am going to refound Western Civilization like St. Stephen did or St. Paul or Daedalus or Socrates did. I am writing my fifth book. I hope to write twenty plus books. The thesis of my book is that Shakespeare had an obsessive-compulsive personality and was anally fixated. <u>A Midsummer Night's Dream</u> represents a try to sublimate his feces. I am going to apply a Rorschach Manual to prove my thesis.

        I went up to Duluth for the last few days to visit Mark McGee and John Herlick. Mark McGee has written ten books on relationships and has hardly any relationships. Surprisingly enough, John Herlick is having problems with depression. He has a doctor and is taking Prozac to help him with his depression. He has put in nearly thirty years at Burlington Northern. He doesn't get much appreciation for his efforts. This would depress me if that happened to me. He might try to get a Ph.D. in psychology from a trial university. He is in a M.A. degree program. I hope that your hand is healing up well. Mother and Dick are returning to Minnesota in early June. Steve Alm is going to build a new garage for Mary Ann. I am going back to my thirtieth high school reunion. I guess that John Kennedy had a 119 I.Q. He got a Ph.D. from Yale with it and published a lot also.

        I was a socialist and I want to end up a radical Catholic republican. I feel that Newman Center is pushing a Catholic guilt trip about everything, that is, racism, sexism, the environment but <u>not</u> sex. Life is problems and suffering so why feel guilty about life? Do you cry over acid rain? All they have at Newman Center is leftover 1960s programs. They have been pushing these programs for twenty-five years. I'm find them boring. They have an Italian priest there who wants to play a power game. Why play the game?

IRISH PSYCHOLOGY/IRISH PSYCHIATRY: THE TIRESIAS COMPLEX

Love, Patrick

Rene, the genie,   6/3/94

The thesis of my book on Shakespeare is that he had an obsessive-compulsive personality which correlates with an anal-fixation. Shakespeare mixes up sex and violence with dreams and poetry in an obsessive-compulsive style. The play chosen to amplify this thesis is <u>A Midsummer Night's Dream</u>. It is a wet dream and an expression of his libido. Shakespeare's preoccupations with sex, violence, poetry and dreams are a lifelong try to sublimate the material or the feces. The <u>Dream</u> is a reflection of this process. It climaxes with Titania, a fairy Queen, falling in love with a man wearing the head of an ass--bestiality. Titania cathects with the ass/man in a dream within a dream.

I brought Mary Ann a rose yesterday for a present at a dinner she had invited me to. Mother and Dick arrived from Florida last night. They are going to a recital tomorrow night, Saturday, for Julie's daughter, Kelly. I am going up to the river on Sunday to visit them. I suppose that you are jumping for joy over the Pope's decision to restrict the priesthood to men only. I might buy the new Catholic Catechism. In his philosophy Ph.D. the Pope related Scheler's personalism to Catholic ethics. Dorothy Day's philosophy was personalism. She related it to the Catholic Worker Movement.

I am going to give you my Ph.D. for your birthday present when you get back in the USA. I should have my book on Shakespeare done by then so you should get a copy of that book also. I would like a copy of your dissertation when you get done with it. I would make it a part of my institute. I want to write around twenty books like Orestes Brownson did. I want to save him from the dustbin of history. I am going to dedicate my short story, "The Classical Father," to Joey Hart who died in Vietnam. I like writing better than teaching.

Write me soon, love,

Patrick

John O'Dougherty and Mike Franey made some suggestions for my Shakespeare profile.

Maureen,   7/20/94

The interview with Dr. Gendron on last Monday went well. I told him that I didn't feel completely comfortable with Dr. Abuzzahab and I preferred to remain with

IRISH PSYCHOLOGY/IRISH PSYCHIATRY: THE TIRESIAS COMPLEX

Dr. Gendron. He agreed. I mentioned that I was feeling very well on my present medications. He didn't think that I needed to change them. He told me to come back in a couple of months.

Aunt Phyl had cancer surgery last week and is recuperating at Methodist Hospital in Rochester. The cancer has metastasized. I went to visit her yesterday. She wasn't feeling well and vomited while I was there. I visited with Eleanor Ann and Gene for a while. Then I went out to lunch with them and mother and Donna and Mary Pat and Kathy. After the lunch, I returned home.

My fifth book is in the final stages and it is my most creative book. Have you ever thought about writing a book on the anthropology of insanity? I bet it is a developing field. I read a book on the history of psychiatry in Minnesota. It was an edited book. However, the author did not mention one psychiatric patient in the history of Minnesota. Do they think we are proletariat? There isn't a psychiatric patient classic in American history. I want to create one. You don't have to share your research with me, but why don't you share some of your ideas with me? Peg and Mary Ann do. Isn't that what an intellectual is all about?

Have you ever thought about getting a literary agent? They must have agents in New York who specialize in third world or anthropology books. Mark McGee has an agent. Mike Franey's girlfriend got an agent and book contract from an agent who specializes in children's books. She is writing up a curriculum called the "Kindness Curriculum" for young children. I prefer writing and research to teaching. A book is going to stand the test of time much better than a test. Campus life is a counter culture, but teaching is frustrating. I saw Brazil win the world soccer cup. Congratulations!

Do you think it is a good idea to remain an apostate Catholic? It is a mistake. Life is short and there are many adversaries in the world. There might be a spiritual adversary. The key to life is simplicity. Single people have written many of the classics. Women have been teachers since the beginning of time. They are weak in research and publishing. Many of the professors in the history department have publishing contracts and agents. There are all kinds of job options for a Fulbright scholar: teaching, publishing, think tank, government and house wife. Keep your options open. My books aren't best sellers, but if you write about twenty books you are in the top one percent of the writers wouldn't you think?

Share some of your ideas, write me soon,

Dr. Patrick

IRISH PSYCHOLOGY/IRISH PSYCHIATRY: THE TIRESIAS COMPLEX

Maureen,   9/23/94

How do you like my new stationary? Mark McGee and Mike Franey helped me with it on my computer. I got an article published in the <u>Minnesota Daily</u> today. It is the one I sent you on Newman Center. Bishop Lawrence Welsh from the St. Paul Chancery called me up on Monday and encouraged me to write him. I have five books published and two articles. I think that I have a very formidable position. I am going to contact some of the leading historians in the country on my efforts to create a Catholic/Jewish synthesis on campus.

I think that American history was often a low spot in the decades before the twentieth century. Slavery, segregation, and the suppression of women make these years a low spot. Give us a new benchmark! Shakespeare was "myriad-minded." My thirtieth high school class reunion is tomorrow.

The woman wins,

Dr. Patrick O'Dougherty

Patricia and Dick,   7/26/94

The problem is not real estate values in Camden. The problem is John Bohanon. Legal advice is <u>not</u> enough. You need good real estate advice. There are many real estate agents in North Minneapolis who specialize in business and residential property up there. All they do is deal with mortgages, contract for deeds, estate sales, forced sales, interest rates and balloon notes in that locale For example, the agent at Edina Real Estate that I talked to thought a person should get $35,000 down from the lawyer, in this case, John Bohanon, and the rest in 3, 5, 7 years on a balloon note. This is one way to finance it. There are many. You should get an income tax statement from John Bohanon for last year to see what he has been doing with the rent from that property. You got one from Leon Trawick. John Bohanon lives at home and doesn't pay any rent. He must have a lot of money. He knows how much you make about. You have a right by law to one half of the rent money. You have a legal right to an accounting on rent money. It is probably a very <u>standard</u> legal and real estate procedure.

Just use good business advice and common sense on this deal. I just made two stops in downtown Minneapolis and one phone call to Edina Realty and got all kinds of information. The Bar Association has an arbitration board to look at legal fees and practices. It is probably free. The Attorney General's office in St. Paul has free advice also. The real estate board has an arbitration board. The banks have one too. The insurance companies have one also. The stock brokers have one. They are all free. They want you to ask their aid. They have these in every state and they can put you in touch with their people in each area. All it takes are a few phone calls.

Dr. Patrick O'Dougherty, businessman

Patricia Kast,

1/10/95

Heart Issues

One of your strengths as a woman is having a strong personality. You need one for social work, politics, and business. However, your problem is you can't accept criticism and you are negative about me. Most counselors tell clients to shut the door on nagging and faultfinding. If you are insecure about academic achievement, the main way for **you** to deal with it is to produce something scholarly, for example, a book, music score, or a work of art. Did any of the Thomas and Emma O'Dougherty family do this? Many people with high I.Q.s never produce anything. Some psychologists give Cervantes and Copernicus 106 I.Q.s. Yours is much higher. My field, intellectual history, is often about ego bashing. The new telephone deal is all right with me. I give a lot to other people. They don't always reciprocate with me. This is probably true of telephone calls. Do you call me often? Thanks for the book by Mother Teresa! I wanted the VCR so I could watch some movies over Christmas. I still don't have it. Who stood up for you and my father in our family?

Patrick A. O'Dougherty, Ph.D.

Patricia Kast,   1/20/95

Time is Money

I have contacted some lawyers in the business about what to do about my father's law office, estate and law practice. They recommended a civil remedy in district court. I am a beneficiary in that estate. I have a right to take this action. I am going to take this advice as soon as possible. You have left this situation hanging over our families heads for ten years now. This is ridiculous. You are a social worker with little bargaining position against a lawyer or lawyers. Mike O'Dougherty and Margaret Wright have children and their money goes to their children. Mary Ann and Maureen know little about business. They don't have much of a fulcrum here. If you do business with friends or family, they are going to want a real deal. Peg, Dick Hope, Mike Franey, Mary Hope, David Noble and Terry Gips have all told me to do this. Dick Kast had nothing to do with that apartment building. However, he has done some other business deals in our family. There are hundreds of people who will or can be executors of an estate. If you don't wrap this situation up with John Bohanon as soon as possible, I am going to take the initiative. **You had your chance here for ten years.** You could reinvest the money. Come to your senses. If you end up in a nursing home, what will John Bohanon give us? You have been remarried for five years now. Let's move on.

Reason not tears,

Patrick A. O'Dougherty, Ph.D.

Patricia and Dick Kast: 1/23/95
A Woman Attorney
    Dr. Frank Preston, my internist, told me that typically these professional relationships will resolve without a **deadline**. Also, he told me the attorneys are not going to give it away. I get the Women's Press here at Riverside Plaza. **I am going to hire a woman attorney and to have this situation resolved.** I have been given this advice. My father, Aquinas O'Dougherty, told me that John Bohanon was the problem partner in the firm. For example, he <u>declined</u> to let Bernie Kukar buy into the office building. I am an heir. I have a perfect right to contest this estate. The problem is John Bohanon is difficult to do business with. He has been given too much slack. Terry Gips told me this type of situation can resolve **rapidly** with an attorney. Everybody has given me the same advice here.

Patrick A. O'Dougherty, Ph.D.

Patricia Kast: 1/25/95
The Downstroke
    I am going to private attorneys by your 76th. birthday, February 10th., 1996, about this John Bohanon and Jack Carleen situation. You have not even gotten interest on that investment for ten years! Aquinas O'Dougherty was my father. I have a perfect right to do this. I am <u>not</u> going to stay out of this deal. You are making a real mistake doing business with just friends involved. They want too much of a deal. Look what happened with John O'Dougherty and Laszlo Zendry(sp.). He talked John O'Dougherty out of his pension. You paid fees on the home, the condo, the apartment building. You should pay a fee on this building situation to keep John Bohanon and Jack Carleen honest. Margaret Conant and Dick Hope all think you should rapidly address this situation. I have been given professional advice about this. I am going to take it. Also, lawyers have a fee arbitration board. One

can go to the bar association too and to their law schools. Have the building refinanced or cut a deal with John Bohanon.

No excuses, the downstroke!

Patrick O'Dougherty, Ph.D.

See enclosed copy of the Women's Press with their listings.

Dear Brother and Sisters,   1/30/95

The Mother Who Kept Her <u>Schizophrenic</u> Son Locked Emotionally in the Basement or Cellar

   The basic problem the counselors or psychologists suggest to me that I have is the insecure, rejecting, intrusive, nagging mother. She keeps her son locked emotionally in the cellar.   Mother wants to get a lot of social approval from these counselors. She wants them to give me a lot of criticism and rejection and they don't. She wants them to clip my wings and tether me. I never married because she has given me a life of constant nagging and fault finding. I could never do that to a child. The reason that she has interest in my problems is because she is the problem. She doesn't like it that she had an intellectual son. I can't help that. She doesn't like most of my friends, my intellectual interests, for example, or the MMPI, or my girlfriends. She is somewhat emotionally inadequate, so she is there for every problem but few of the successes. A lot of women would give their teeth for an intellectual, good-looking, personable, interesting, achievement orient son. However, she just wants to undermine constantly my male self-esteem.   Some girls have fathers who want to keep them down, but in my case it is the mother. I don't think there is anybody in our family that would disagree with this. I have always had pretty normal personality profiles.

This is the professional opinion,

Patrick O'Dougherty, Ph.D.

Maureen,   2/11/95

Faust

IRISH PSYCHOLOGY/IRISH PSYCHIATRY: THE TIRESIAS COMPLEX

Faust found happiness in life when he got a good job as a scientist. I hope that you like your job. Thanks for the letter. I'm glad that you like Audre Lorde. She tried to be formidable. So does Kate Millett. Relationships are a big part of life and so is sex. However, I think many women make a mistake by slipping into a life of carnality. Sex can be "very distracting." This is a point Edith Stein makes. Mary Ann is producing a compact disc. If you don't have a business in life or produce a book or recording, you are going to end up with nothing. There is a premium on intellectual contributions in Western Civilization. I've made a lot, so I'm not going to have any trouble. So has Peg. I hope that you do also. Talk is cheap. Produce something. Get your students to produce something. I am sending you my "Austrian Studies Newsletter." It contains some real first rate research in history. Most people are "easily replaceable integers." Writers and intellectuals are not. I'm going to devote the rest of my life to writing.

Love,

Patrick A. O'Dougherty, Ph.D.

I got a mock Nobel Prize.

Steve, Jeri, Terry,

4/3/95

I talked with Dennis Cummings who has an M.B.A. in business administration from the College of St. Thomas about the administration problem at Newman. He told me (1) ask for yes or no answers, (2) if there is a roadblock go around it, (3) file actions, for example, to move IASA and its library or to change the staff at Newman, (4) don't get involved in situations where you are spinning you wheels, (5) I personally think there is a premium on intellectual and artistic contributions in Western Civilization about promotions, (6) what are they going to do for Patrick O'Dougherty at Newman? Why have so many people left? Is it a Christ versus Caesar problem? Dennis doesn't think a monopoly is the way to administer a business or a center. The best committee is "a committee of one."

Patrick A. O'Dougherty, Ph.D.

Maureen,

6/3/95

The Mark of the Christian

The defining mark of the Christian, the Aryan, or the Caucasian is to live for the good--to live in sacrifice for the good. Animals do not or cannot do this. The best people, for example, Christians, Jews, Aryans or Caucasians live for the good.

IRISH PSYCHOLOGY/IRISH PSYCHIATRY: THE TIRESIAS COMPLEX

For example, many other civilizations do not have missionaries. The whole focus of Catholic theology is the common good. Without sacrificial love, there is little excellence. The cross represents sacrificial love. So does the Blessed Mother. The Japanese did not have the cross.

Personalism

Thanks for the reply. My sixth book is almost done. It is on personalism and mathematics as women's personifestos. Personalism is the philosophy of Dorothy Day and Karol Wojtyla (Pope John Paul II). It looks to the centrality, rights, social obligations, responsibilities, and personality of the person.

The Mother

Patricia Coyne is a good mother, but she is a critical mother for me anyway. She doesn't want to let her children go. John Herlick thinks that this is true for many women and for Patricia. I am a single man and I have my freedom. I don't want to stay in the yard. I am not a husband who the wife doesn't want to stray. Mark McGee thinks that Patricia wants to keep me locked in the basement. If you get married and have children, you lose your freedom. Clarke Chambers who was a professor at the University of Minnesota told me "not to knuckle under in life." An intellectual should not "jump on a merry-go-round" thinks John O'Dougherty. I am having problems with the mother. The doctor she worked for should not resolve them. It would be unprofessional. I had **major surgery** and tardive dyskinesia with Dr. Gendron. The woman lawyer at the Attorney General's office thought I should file this. So does John O'Dougherty, and Ken Conant. I can get all types of free legal advice from many attorneys in the state who represent psychology consumers. David Noble's secretary told me that the medical profession is a "house of horrors." Medical malpractice is very common. If you have a doctor who has to repeat a surgery, you should file it. Many doctors have bad litigation and bad medical records.
Patrick A. O'Dougherty, Ph.D. **Mother is losing almost $1,000 a month in interest and principal on the office deal.**

Maureen,   7/8/95

I wonder if you could send me some anthropology references on distilling cultural truths. I have a book here that Peg gave me for my birthday by Kate McGowan and Anthony Easthope called, A Critical and Cultural Theory Reader. It looks at the "'death of literature' and the rise of post-structuralist theory [which] has breached the traditional opposition between the literary canon and popular culture." The book is intriguing because it does not have empirical studies. A theme from

IRISH PSYCHOLOGY/IRISH PSYCHIATRY: THE TIRESIAS COMPLEX

Catholicism is "triumphalism" which is purifying Catholic cultural truths. Good luck on your research.

Patrick A. O'Dougherty, Ph.D.

Sean Wright,   7/15/96
Missed Opportunities
What I was trying to have a talk with you about at Christmas time is the <u>downside</u> of good looks and too early sex. Maybe you are too young for a talk about this, however, your father must be around the age of seventy by now. You should hear about this. Don't get involved in a teenage pregnancy! Don't get married too young! Don't have an abortion! Don't lead a deviant lifestyle, for example, a homosexual lifestyle. You will miss out on all kinds of opportunities in life if you make these errors. The strength of a Catholic education is that the Church doesn't encourage the young to wreck their lives. Don't wreck yours!

Great Grandmother, Margaret Coyne
Margaret Coyne, your great grandmother, told me just before she died that I should never stop my thirst for reading. I give you the same advice. Don't ever stop reading, writing, and doing mathematics.

A Writer's Lot
Terence O'Dougherty, your great uncle, thought that what I should do if I wanted to be a writer is to "not do experts and textbooks." I should just walk out the door after college with little money and whatever happens to me in life happens. I did that. My father, Aquinas, liked the book, <u>You Can't Go Home Again</u>, by Thomas Wolfe. He thought I should use that as a tool to develop my writing experiences. Neither Terence or Aquinas wanted to be writers. They gave me the right advice though don't you think?

Your Uncle, Patrick: A Little Unconventional
Unlike your mother and your father, your uncle, Patrick, is a little unconventional, a little eccentric. However, that is his strength as a person and as a thinker. Don't you think?

God Bless,

Patrick A. O'Dougherty, Ph.D.

Nellie Bolie O'Dougherty
963 Demont Ave.

IRISH PSYCHOLOGY/IRISH PSYCHIATRY: THE TIRESIAS COMPLEX

Maplewood, MN. 55109

Nellie,

11/14/95
The "Immigrant" Story

There are immigrants and "immigrants" to America. Nellie Bolie O'Dougherty is our immigrant from Greece to America, right? So is her sister, Katerina, right? I'm not trying to start a big dispute about this in our family. However, think about it! This is a big problem in immigration perception, analysis and history.

Patrick A. O'Dougherty

Margaret and Frank Wright: 12/14/95

The Pocket Nerve

I hit the pocket nerve in the Wright family with my challenge of Mario Cuomo's liberalism. Most of the medical grants in America are liberal grants. Mario Cuomo had a very difficult time with many of the WASPS and other white people in New York. I don't know why he did. The women in this country want half of the positions in most fields of endeavor. Why not Catholic equity in the secular atheistic/agnostic school system--one fifth of the positions? The Catholics bring more than that in trade into this country in trade. It might be as much as half the foreign trade. Most of the promotions in the secular universities base themselves on the betrayal of Christianity--dollars for dogmas. How many Irish Catholic males have they ever had on the faculty of the Ohio University system?

Salvation

The salvation of your soul is a personal and collective responsibility. It is up to you what you want to do with this responsibility. All I can do is to pray for you.

Graham Greene
A Catholic Convert and A Favorite of Aquinas O'Dougherty

"If I had to choose between life in the Soviet Union and life in the U.S.A., I would certainly choose the Soviet Union."

Attributed to GRAHAM GREENE--Parade magazine, October, 29, 1967, p.2. Unverified.

I wrote my seniors honors thesis on the convergence of American and Soviet thinking and cultures. It is getting more true everyday.

The Pocket Nerve

77

IRISH PSYCHOLOGY/IRISH PSYCHIATRY: THE TIRESIAS COMPLEX

What did Margaret, Mary Ann or Maureen ever do for me financially? I am the main person who did the apartment deal. I hope Maureen and Mary Ann come off on their Latin American adventures. I paid a terrible price in life to get where I am. The young people today want to start at the top. What did they do?

Dialogue
12/14/95
Psychology and Medical Research

The founding on this date of the personalist/intuitionist school psychology and medicine by Patrick A. O'Dougherty, Ph.D. The founding of the hydrogen conversion and alternatives to petro-chemicals school of medicine, especially pediatric neurology on this date by Patrick A. O'Dougherty, Ph.D.

Let me know how you come out on Mario Cuomo, love

Patrick A. O'Dougherty, Ph.D.

Margaret and Frank: 12/15/95

A Loving Farewell to Liberalism

(Paraphrasing Yeats "A Loving Farewell to Christianity")
The DFL politely asked Patrick to leave. Paraphrasing Robert Frost, two roads diverged in midpath, I took the **higher one** and that made all the difference. I bid the DFL a loving farewell.

The Single Majority

I don't want to be intrusive in your family. I would never do anything to be a bad influence on your children. However, what do married people or children care about a single person like me? I've always worked. I have a formidable research project on campus.

Dialogue,

Patrick A. O'Dougherty, Ph.D.
Mary Ann gave me her digital for my birthday.
In the last letter one-fifth needs a hyphen

Margaret and Frank: 12/18/95

78

IRISH PSYCHOLOGY/IRISH PSYCHIATRY: THE TIRESIAS COMPLEX

Mother told me you wanted an apology. Here is a written apology to you. I like the Republican Catholic tradition of Orestes Brownson, Archbishop John Ireland, Cardinal Dennis Dougherty, and William Buckley and Patricia Buckley. I don't feel comfortable with Mario Cuomo and Ted Kennedy. Catholics are one-fifth of the population in America. They account for more than half the foreign trade. Did you get your fair share? I don't want a big conflict over Christmas.

Love,

Patrick A. O'Dougherty, Ph.D.

Margaret and Frank Wright: 12/27/95

Confidential: Just Between Margaret and Frank and Patrick

I asked at the Attorney General's Office about what to do if two lawyers, like John Bohanon and Jack Carleen, tie up an estate without compensation for an excessively long period. They recommended the Office of Lawyers Professional Responsibility to me. Three of the women at that office recommended filing a complaint. I am sending you a copy of the pamphlet they sent me. I am also sending you a copy of the complaint I filed. Dick Hope, Mike Franey, my internist, and an unnamed woman psychologist also recommended this action. Anyone can do this. I will let you know the outcome of this action.

Patrick A. O'Dougherty, Ph.D.

Patricia Coyne,   2/1/96

Patricia Coyne and the Clutch

Who do you have in the clutch? Why don't Margaret, Mary Ann or Maureen say something about the law office building for you? You always play yourself down; and, you always play me down. That is the problem. I'm the main one who speaks up for you in our family. We have incurred losses in our family. They are John Bohanon and Jack Carleen. The loss is time. They have our families' heads on the guillotine.

IRISH PSYCHOLOGY/IRISH PSYCHIATRY: THE TIRESIAS COMPLEX

Patrick A. O'Dougherty, Ph.D.

Points in My Mother's Favor

My mother has always been tough on me, but a point in her favor is that she has been there for me in the clutch. Maybe the reason she doesn't play herself up is that she doesn't have to.

Margaret Wright,   2/6/96

A Psyche Consumer: Major Surgery

I have been a psyche consumer since the late 1960s, like many of the people in my generation. However, I ended up with major arthroscopic knee surgery five years ago at St. Mary's Hospital due to Prolixin and Thorazine over medication. I suffered terribly for many years with tardive dyskinesia which created Parkinson like symptoms among many other symptoms. It is also maximum daily discomfort which went uncompensated. Also, I have always worked since I was eleven years old. You sought counseling from Dr. Loper. What would you do if this happened to you? Think about it.

The Rejection Card

Most psyche patients are less violent than the public. They are also typically the victims of family rejection or abuse. This is true in my case. For example, I never took boxing, or karate, or played much football, or went into the military. I am single person who has never married, and if I have a problem I typically dealt the rejection card, especially by my mother Patricia, who does not feel comfortable with a very creative child. It is neurotic guilt and rejection on her part. My sister, Margaret, has many strengths. However, she is quick to take offense and one counselor told this is a difficult type of personality to get along with. Margaret told me she has a paranoid personality. For example, she took offense at Rudy Boschwitz, my father's law partners, and the Germans. My sister, Maureen, is also rather cool and negative with me. Rather than deal with my problems or my issues, they refer me to counseling. My family has almost <u>never</u> been there for any of my <u>many</u> successes: my graduation from college, my M.A., my success with our family apartment business, my run for office, my election to the head of the residence council at Guild Hall, my published Ph.D., the Legion of Mary, most of my books, my highly recognized library, my archive, and my legal cases. They are just there for my problems and my few faults. I don't really know why my family deals me the rejection card. It is to personal insecurity on their part. I am the one who has done a lot for them. For example, I have gone to many of the concerts of my sister, Mary Ann. I asked my sister, Margaret, why she wants to run me down. I asked you not to do this.

Patrick's Place in the Sun

IRISH PSYCHOLOGY/IRISH PSYCHIATRY: THE TIRESIAS COMPLEX

I tried to stand up for my father's legacy by addressing his law office estate in a suit with the Office of Lawyers Professional Responsibility. The Italian priest at Newman Center, Father Bossi, is too Bossy. I just tried to stand up for my interests in a subtly negative relationship by suggesting an amicable lawsuit. This prompted the evaluation. I want to stand up for my lot in life "from where the sun now stands." This is the title of a book loved by my father.

Irish Issues

I come from a large hopefully loving Irish Catholic family. The Irish can't build a large aircraft carrier out of the peat bogs in Ireland. So people like the Irish and others think that what they should do to keep themselves percolating is to take up the Celtic Cross and to brace the Semitic and black realities. I hope that my family heads this advice.

The Birthday Gift

A counselor at HCMC didn't think that I should have returned the Mario Cuomo book that you gave me for my birthday. However, I told her that I just asked for another book and that I didn't feel comfortable with the book on Woody Allen and the one by Mario Cuomo. Taste is a big factor in books. My sister, Margaret, is quick to take offense at anything that isn't liberal, for example. What I am going to do about this returned gift is to apologize for it?

On Patrick's Ph.D.

I did not get squeezed out of Ph.D. When I was in graduate school at the University of Minnesota, I did not want to take up the yoke at a young age like you did with Keith Nuechterlein. I wanted my freedom. My committee agreed to leave the door open for me at the University of Minnesota. I did my dissertation with David Noble, Al Nier, and Mike Franey who I felt comfortable with. David Noble made the history of the University of Minnesota and Al Nier made the history of science. Almost everything is up for grabs in education. I have invented a new school of Catholic philosophy, personalist intuitionism. I'm pro-life all species. I have published seven books. I have a major research library that I have built at Newman Center at the University of Minnesota. I am founder of the alternatives to petrochemical movement which will revolutionize most fields of medicine. I have an archive which I hope will become a Minnesota psychiatric archive. What do you have as a thinker? Did anyone ever raise the lightweight issue with you?

Good Looks and Sexuality

The reason that I raised the issue of good looks and sexuality with your son, Sean, is because his father, Frank, is in his seventies. We ended up with this Jacob Wetterling incident in the small town of St. Joseph, Minnesota. Think about it. I was trying to warn him of the downside of good looks. Just look at all the milk carton kids out there. I have always been a backer of your boyfriends, husbands and children.

Intellectual History

Intellectual history is about volleys and challenges and is a quite provocative perhaps personally offensive field. For example, I'm in recovery from the South. My threats have been intellectual or legal only.

God Bless and love,

Patrick A. O'Dougherty, Ph.D.

Patricia Kast,   2/15/96

A Quite Positive Portrait of My Mother
A Catholic Intellectual Portrait

Catholic theology is a science. I did a scientific analysis of your personality. On the MF scale you are feminine. Your other high scales are hysteria and paranoia. You are a postcolonial women's thinker/activist, for example Hubert Humphrey's Vice DFL chairwoman.   And you are a scientist. Social Work is a science, like Catholic theology. The key to your life is Benedictine hospitality. It was a key to the life of Dorothy Day who involved her movement at St. Benedict's College where you went to school. Like Dorothy Day, a mentor of yours, you are a Catholic Worker, in this case social work. Your mark as a woman is purity. This made you a good role model for working with prostitutes and American Indian mothers. You don't smoke, drink or swear. You like to read and have a good eye for detail which is a mark of intelligence. You worked at Hartford Mental Retreat for Dr. Ted Liebermann who was a famous psychiatrist.   The cottage plan for psychiatric care came out of Hartford Mental Retreat. In counterpoint we had a cottage on White Bear Lake. This was the cottage plan in our family for child rearing. You are a friend of Senator Eugene McCarthy the peace candidate.   You have businesses in the black community. Your theology is liberalism.
Also, you are a writer.

Love,

Patrick A. O'Dougherty, Ph.D.

Dick Kast,   2/16/96

A Single Person Out on a Limb

IRISH PSYCHOLOGY/IRISH PSYCHIATRY: THE TIRESIAS COMPLEX

I am writing a formal apology to you for a criticism or slight I made to you about business issues in our family. It is not easy on relationships to have a business or businesses in a family. Thanks for bringing up your problem with this criticism. Also, thanks for bringing up the phone bill problem. I am the type of person who starts calls with people rather than have them call me. Why let people lean on me or us? Also, I don't want to make calls down to Florida about money issues all the time. You have been a good enough guy to me and our family.

The deal is I don't want to be out on a limb with lawyers and business deals if my mother who is seventy-six dies. I have to look out for myself. I am enclosing a letter I sent to John Bohanon and John Carleen about stealing the rents.

The Green, the Grey and the Purple Revolutions
I like the grey and purple color dimensions you add to our family. They are musical dimensions. I don't know a lot about accounting, music and mechanical tools like you do. So, the relationship is complementary. I also like the Protestant relationship. You are a good Christian man.

Praxis
It takes years of practice to become a musician. Did you get paid for practice? Writers are in the same sort of situation. How many people would read all the books I have? I chose the idea of the wanderer for developing my intellectual content. It is a good one.

Big Price, Big Prize, Love,

Patrick A. O'Dougherty, Ph.D.

Dr. Frank Wright: 2/21/96
Medicine Noir
In honor of your upcoming birthday I have invented the fields of psychology and psychiatry noir. Noir is French for black and the word also shows "the black numbers in roulette." This is what Patrick O'Dougherty is all about noir and chance factors in writing and life. I was a "slave" in the South in Atlanta, Georgia, for eight months. St. Patrick was also a slave.

Alternatives to Petrochemicals
I am sending Sean and Megan their belated Christmas gifts. Can they figure out some alternatives to petrochemicals in these products. The alternatives to petrochemicals area of research is a potent area in medicine today.

IRISH PSYCHOLOGY/IRISH PSYCHIATRY: THE TIRESIAS COMPLEX

I would appreciate a thank-you note. I don't know if I will get one right away. Good luck backing Mario Cuomo! You will need it.
Happy Birthday,

Patrick A. O'Dougherty, Ph.D., wanderer

David,   3/1/96
Patrick A. O'Dougherty: Up from Slavery in the South
Here is my new Ph.D. thesis for you my phone number: (612) 339-1748. I love St. Patrick. I took his defeat by becoming a slave in the Deep South for my intellectual trip. I want to build a scientific revolution out of it. Do you like St. Augustine? He had a child by a slave woman. Out of this experience, he invented systematic theology in Western Civilization. These are the types of role models the students at the University of Minnesota should hear about. My sister's digital sold 100 copies opening night. I hope you like it. I never have dated much in my life. I have always gotten invitations from women. However, I think it is an intellectual mistake to lose your life in carnality. I don't advocate a lot of guilt about sexuality. However, I think a teenage pregnancy or getting married too young can be one of the dumbest moves you can make.

Patrick A. O'Dougherty, Ph.D.

Maureen: 4/1/96
The Green Revolution
Happy Easter! For your Easter gift, I am sending you some information on the Constitution for the Federation of Earth and the World Parliament. I am for the Green Party, Green Caucus internationally. Petra Kelly founded the Green Party in Germany. Back the Green Revolution! I am also sending you some information on silver mercury fillings which are a health hazard. They are illegal in Europe.

Patrick A. O'Dougherty, Ph.D., revolutionist
"'Politics is the shadow big business casts on society.'" John Dewey

A Catholic, Scientific Conceptual Framework and a Methodology
There has been a massive intellectual contribution to mathematics and science made in the twentieth century. Catholics should create a new Catholic scientific revolution. For example, one can take a subdivision of a field, like semiology the analysis of signs in linguistics, and apply it scientifically using a methodological technique to analyze signs in biblical texts or in Catholic theology. This can be done for thousands of new areas in research in all levels of science and mathematics. The

idea is to invent new fields, like the personal intuitionist school of physics, and apply them to theology in a manner consistent with Catholic teaching.

Margaret Mary, Mary Ann, Maureen,

4/8/96
Men
Edith Stein thought the downside of men was that they are brutes, slaves to their work, and single-minded.

Women
The downside of women is that most of them are quite "defenseless," and they are for often more "devious" than men, especially over pregnancies, and infighting. They are more prone to slip into a life of carnality than men.

France
"France is the Church's eldest daughter and her most prodigal."
Yours in Christ,

Patrick A. O'Dougherty, Ph.D., The Green, the Grey and the Purple Revolutions

Patricia Coyne: 4/29/96
The Burden of Status Problem
The oldest child often has the burden of status in a family. One psychologist told me the reason for the nagging and faultfinding from the mother is she had me take the pressure on the burden of status this way. Mike had a big falling out with his father before he died. This might be somewhat normal. Do you want to have a big falling out with me before you die? Request that in writing if that is what you want. Why didn't you address the relationship issues that I wrote you about? You know little about psychotropic drugs. I have asked you to quit nagging me about them. Drop it. The medical profession is replete with horror stories. I found out about it. So did my father. So did Terrence.

Patrick A. O'Dougherty, Ph.D.

Patricia Coyne: 5/1/96
Family Matters
I talked to a Catholic woman psychologist about family matters. She thinks the reason you are critical and faultfinding with me is because the oldest has the burden of status. She thinks the deal with Mike is that he didn't pick particularly

85

good wives. They are very dominating. She suggested that friends and intellectual relations often move on. However, your family relates genetically to you, and they should grow with you. She thinks that I have a pretty good relationship with Mike O'Dougherty. I should meet him for lunch every once in a while. She thinks that I should intellectually send the Greek wife back to Greece and that I cannot do much about her. Don't you think that it is a pretty big intellectual trophy to make the World Shakespeare Bibliography? It happened on Earth Day. Mark McGee taught at Texas A & M. He was ecstatic about this. The Coyne woman on the Minnesota Supreme Court gave her chair to Kathleen Blatz, a German woman. I think Dick Kast is a good choice. Why try another Irish Catholic in our family? I wish I had his musical and practical talents. I'm also for the black and grey realities believe me. That is what keeps an Irish Catholic clean. I will try to get a hold of this women's directory for your mother's day gift.

Love,

Patrick

Patricia Coyne: 2/15/96
A Catholic Intellectual Analysis of Patricia Coyne
Catholic theology is a science with a mathematical basis. On the MF scale you have a very feminine personality. Hysteria and Paranoia are also prominent scales in the scientific analysis of your personality. There are powerful themes one can use to analyze your life. You were a social worker. Social work is a science. Your writings are postcolonial writings. You were a postcolonial political activist as DFL Vice Chairwoman with Hubert Humphrey. Dorothy Day was a mentor of yours; and, she was a Catholic Worker. You were a Catholic social worker. Like Dorothy Day your mark as a product of a Benedictine College is hospitality. Another mark of yours is purity. You have a good eye for detail which is a sign of intelligence. Senator Gene McCarthy is a personal peace candidate friend of yours. You worked at Hartford Mental Retreat with Dr. Ted Liebermann. Hartford Mental Retreat introduced the cottage system of psychiatric care. We lived in a cottage on White Bear Lake. This is a parallel in all of our lives with the cottage system at Hartford Mental Retreat. You have a Catholic liberal theological bent. You married a man, Aquinas, named after a great theologian.

Margaret Wright: 6/27/96

Margaret Wright: "Through a Glass Darkly," St. Paul 1st Epistle to the Corinthians 13:1-13

IRISH PSYCHOLOGY/IRISH PSYCHIATRY: THE TIRESIAS COMPLEX

Why do you always focus on the negatives so much: your book, your study in Southeast Asia, Patrick, John Bohanon, Keith, and Rudy Boschwitz? Why can't anybody disagree with you? On scandals, "you can get something on anybody." I have never even watched a soap opera all the way through. I don't read scandal magazines. I don't want to confess my sins to the public. I confess them to the priest. I want to focus on the positive and move on in life.   If I have done something to offend you, I apologize for it. The keys to Margaret Wright are the above passages from St. Paul. Please read them.

Happy Birthday, love,

Patrick A. O'Dougherty, Ph.D.

Timothy,   7/1/96

Terence O'Dougherty and Women
Catholics are giving in on everything over women: poverty, chastity, obedience, and the military. When we lived over on 4104 Bryant when I was four years old Terence lived with us. He was in the Chemistry Department at the time and was having some psychology problems. They were mild. Anyway, he got a few insulin shock treatments at the VA over them. He <u>changed</u> noticeably right immediately. He got a job with the city rather than continue as a chemistry graduate student. He didn't like teaching, so graduate school was not the best option for him. He had a 168 I.Q. on the military intelligence tests. So, intelligence was not the problem. Women were also an issue. Everybody knows about this except your family. Terry was somewhat secretive so he chose not to reveal it. He was a great man and everybody loved him. He thought that what I should do if I wanted to be a writer is to "not do experts and textbooks." I should just walk out the door with little money and whatever happens to me in life happens. I did that. My father, Aquinas liked the book, <u>You Can't Go Home Again</u>, by Thomas Wolfe. He thought I should use that as a tool to develop my writing experiences. Neither Terence or Aquinas wanted to be writers. They gave me the right advice though don't you think? My position is that I don't want anymore rough experiences over women after my experiences in the South. Would you? What are you giving in on over women? You can share this with whoever you chose. Nicole Brown Simpson was a Catholic.

Patrick A. O'Dougherty, Ph.D.
corrected copy

IRISH PSYCHOLOGY/IRISH PSYCHIATRY: THE TIRESIAS COMPLEX

Margaret Wright: 7/12/96
A New School of Thought
I developed a new philosophy and psychology, personalist intuitionism, out of my Viktor Frankl and my wandering experiences in life. I gave a talk about it at the Rajkowskis' wedding anniversary at the Great Hall at St. John's University. Read about it in my book, <u>Personalism and Mathematics as Women's Personifestoes: Women And The Fior Which Is Irish For Truth</u>. This book also grew out of my two year teaching experiences with Cheryl Mandy in the Trio Program in General College. Terry O'Dougherty encouraged me to choose wandering as my writing construct.
Family Matters
I am going to send, first, Sean a personal apology for something that I said that offended him. I am also going to give the twins, <u>Lilies of the Field</u>, made after the life of their grandfather, Aquinas, for their birthday.   Second, I don't feel particularly comfortable with the mother. She nagged and did faultfinding with me constantly for most of my life. She never takes my side on any issue. She has seldom been there for any of my successes. She never plays herself up; so, she has almost never given me any compliments or played me up to any of our friends or relatives. Third, I don't feel comfortable with Mike and Nellie. My psychologist thinks that he didn't choose very good wives. They are very dominating and cool. She thinks that I should intellectually send Nellie back to Greece. John O'Dougherty thinks with them "distance lends enchantment to the view." What do you think? They invite me over there <u>with</u> other relatives once a year. Why bother with them if this is the case? Nellie has never worked one day in her life and doesn't even drive the family car for errands. I don't need involvement with Stefan. This is a wise choice for me. Many of the O'Dougherty nieces and nephews don't have anything to do with their aunts and uncles. What do you think? Fourth, on Margaret, I like her and I don't have a real personal problem with her except that she is often negative and one can't easily disagree with her. She is quick to take offense. Fifth, Mary Ann and Steve just want to be alone. This is okay with me. However, I was Mary Ann's biggest fan. Do you think that she reciprocated well? I hope this marriage works out. Sixth, let's focus on the positive. God Bless,

Patrick

Patricia,
Margaret, Mary Ann, and Maureen,

7/25/96
Displacement
Is the reason my sisters are negative about me do to psychoanalytic displacement?

88

IRISH PSYCHOLOGY/IRISH PSYCHIATRY: THE TIRESIAS COMPLEX

The Eldest Child: The Burden of Status

Oftentimes the oldest child has the burden of status in a family. According to Joyce Forsgren at HCMC, the middle children are the "lost children" and the youngest has to "catch up." Is sibling rivalry a problem here? I don't know. She also thinks that no one leaves "childhood and adolescence in one piece." Also no one or almost no one "leaves childhood and adolescence permanently damaged." Are these problems in our family? Also, I'm glad I didn't go into politics.

My father thought that if he was easy on me, I wouldn't make it. So he got me toughened up with some Viktor Frankl experiences in the Deep South. I was game for them when I was younger. My thinking has gone through three stages (1) changing intellectual thought, (2) Catholic personalist intuitionism (3) and now integration of analysis. How does one apply my school of thought to psychology, psychiatry, anthropology and postcolonial music? Liturgy is therapy and therapy is liturgy. Suggestions?
Emphasize the positive,

Patrick A. O'Dougherty, Ph.D. Irish white Negro
P.S. I have developed a formidable synthesis in Margaret's, Mary Ann's and Maureen's fields.

Margaret: 7/31/96

A Request
I want my sins to stay with my confessor, my psychological problems with the therapists, and my love to stay with Margaret.

Intellectual Courage
What is wrong with a person like your brother, Patrick, who wants to do intellectual courage as well as intellectual preening? I took strong stands on Vietnam, pornography, the white Negro, and Catholic issues. These acts of intellectual courage are my mark as a person. I choose to live at the edge of society. I'm a scientific revolutionist. My psychological experiences have been white Negro experiences. My family has given me an injustice my psychologist thinks in this Newman Center brouhaha. How can I get justice?

Keith
Your father and mother didn't click so well with Keith. However, I liked him. I sent him my listings and I'm going to send him a copy of my research library. I think that you did all right with the Germans. I think Keith did well by us. He liked my "black classic" idea. Do you think that he did any better than I did?

IRISH PSYCHOLOGY/IRISH PSYCHIATRY: THE TIRESIAS COMPLEX

John Bohanon

My psychologist thinks you should address John Bohanon and the business deal.

God Bless,

Patrick A. O'Dougherty, Ph.D.

Timothy O'Dougherty,

The Personalist School of Mathematics

You were not very old when your father died. However, Terrence was a great man. He influenced me to get involved with psyche consuming, mathematics and the Knights of Columbus. He saw to it that I had some Viktor Frankl experiences in the South to turn me into a great writer. I invented the personalist school of mathematics out of these experiences and the personalist/intuitionist school of physics. He had a big impact on our family intellectually. I hope that you come up to his level and the level of all of the O'Doughertys. Did you ever think about becoming a writer? If you do, take your father's advice: Walk out the door with little money and you will definitely be one. You don't have a wife and children what will be your legacy?

God Bless, test of time,

Patrick, mock Nobel laureate

Sean and Megan,   8/23/96

Sean, I've decided to become a revolutionist and vote for the Green Party.

Megan, I'm sending you a copy of The Minnesota Women's Press which has an article on St. Catherine's Summer Algebra Institute for Girls.

In search of personal validity,

Patrick A. O'Dougherty, Ph.D.

Sean and Megan: 9/5/96

Happy Birthday

Sean: **You Only Go Around Once**

IRISH PSYCHOLOGY/IRISH PSYCHIATRY: THE TIRESIAS COMPLEX

I wanted to be a writer since I was about four years old. I wanted to have many different experiences to amplify by writing. I am sending you and Megan a list called "All the Tough Numbers" about some of the experiences and some of the actions I have done in my life. I hope you find it interesting.

Megan: **"Terrible Beauty, Terrible Price"**
You area good looking and intelligent young girl. My philosophy of women is "terrible beauty, terrible price."

Sean and Megan: **Aquinas**

For your birthday I am giving you a movie made about the life of your grandfather who you never met. It is "Lilies of the Field." He worked as a handyman along with his brother Loyola in 1937 for some immigrant German nuns in Colorado. Sidney Poitier, a black man, stars in the role of your grandfather in the film. I hope that you like it.

Patrick A. O'Dougherty, Ph.D., an Irish white Negro in the tradition of your grandfather.

Sean and Megan,   8/23/96

Sean, I've decided to become a revolutionist and vote for the Green Party.

Megan, I'm sending you a copy of The Minnesota Women's Press which has an article on St. Catherine's Summer Algebra Institute for Girls.
In search of personal validity,

Patrick A. O'Dougherty, Ph.D.

Dr. Frank Wright,   10/15/96

Viktor Frankl Experiences Out of Schizophrenia
The Green Revolution

I am a very moral man. However, I ended up getting a genetic disease that runs in our family--schizophrenia. It is not do to a character defect or due to bad child rearing. Schizophrenia is a chemical imbalance due to variation in levels of dopamine in the brain. I ended up having years of personal Viktor Frankl experiences out of it. Viktor Frankl was a Jew who survived concentration camp experiences and went on to develop a new philosophy, Logotherapy, to deal with his experiences.

91

IRISH PSYCHOLOGY/IRISH PSYCHIATRY: THE TIRESIAS COMPLEX

Like Viktor Frankl, I have developed a new philosophy--Catholic personalist intuitionism. Moreover, I am the founder of the Green Revolution on the campus of the University of Minnesota. I think that Minnesota should develop a Psychiatric Archive to relate the problems and stories of people like me. Generational research is necessary on this medical disease. I want to release my medical records to further this research. My psychologist, Dr. Jane Rozsnafszky, thinks your wife, my sister Margaret, is critical, negative and achieving and too tough on her brother. Her sisters and mother are the same way. They should recognize my pain. Also, despite great suffering I have made many contributions in many areas: spiritual, intellectual, political, ecological, and to our family. I think they should recognize these contributions. I am sending you and Margaret my book on the Green Revolution. Why don't you rate it professionally? I am glad that I stood up for my parents and my whole family.

God Bless,

Patrick A. O'Dougherty, Ph.D.

Father Loyola,
6/19/96
Juneteenth: June 19th
Happy Birthday! You were born on a very famous day in Afro-American History: Juneteenth, the 19th of June. It is the day long after the Emancipation Proclamation that black slaves in Texas first heard of their freedom after the finish of the Civil War. The black people celebrate this belated good news. You brought, as a priest, the message of Christ's spiritual freedom to the Southwest.
Liturgy Noir
The 1000th Year of Austrian History
In honor of the 1000th year of Austrian History, I am going to write a major work in liturgy. I have been in the Austrian Studies Program for eleven years at the University of Minnesota. My mother, Patricia, got me involved in this program. One of the themes that I am going to incorporate into this work on liturgy is that liturgy is a home for the displaced, like the psychologically or neurologically displaced. I am going to call this work Liturgy Noir which is the French word for black. It also shows the black numbers on the roulette wheel or chance factors in life and liturgy. I found some books on liturgy at St. Patrick's Guild on the history of the Mass, language and liturgy, issues in liturgy and the church cycle. I am psychologically displaced so I want to include my case study as part of inclusiveness in liturgy. Do you have any ideas on liturgy from the Chicano perspective that I might include? Many Chicanos are displaced persons and need a home for the displaced: liturgy. Another theme

92

that I am going to use is from Virgilio Elizondo--"'Mestizaje: A New Imaging of Place and Belonging'" in liturgy. "'Mestizaje' refers to the mixture of culture and races, of immigrants and non-immigrants. It is the process through which two different peoples mix biologically and culturally so that a new people begins to emerge." Mestizaje is a big part of the experience of the Mexican-American community of which you are a part.

It is a gift to have a priest in our family, Patrick O'Dougherty, Ph.D.

Patricia, Margaret, Mary Ann, Maureen,
9/24/96
F. Scott Fitzgerald
Mark McGee and I went to the F. Scott Fitzgerald 100th birthday party today at the Landmark Center. Garrison Keillor signed my program. The committee issued a stamp in memory of Fitzgerald. There were speeches by Garrison Keillor and Mayor Norm Coleman.
Dr. Jane Rozsnafszky
Dr. Rozsnafszky thinks that our family personality traits are critical, negative and achieving. She is right especially for myself and the women in the family. I asked Stacie Joncas a counselor at HCMC what she would do if she had a mother and sisters like this. She told me to just accept them the way they are. Don't internalize the critical and negative aspects of their personalities, and live your own life. I will give you an example of this. I had a justice issue about Newman Center in my life. Not one member of our family put in a single good word about me. Dr. Rozsnafszky thinks this is what my problem is. My family focuses on my disability rather than my ability. They give me almost no validation for any of my strengths or accomplishments. How would Margaret, Mary Ann or Maureen like it if I received a call about them from a doctor or counselor, and I just revealed their faults or problems to the professional? The main reason I go to counselors is to get away from the criticism, negativity and faultfinding of the women in our family. In contrast, the father usually encouraged me in my endeavors though he used to debate with me a lot. What can be done about this problem in our family?

God Bless,

Patrick A. O'Dougherty, Ph.D.

Dear Margaret,
11/5/96
Would you please send me a positive letter back that expresses understanding for what I've been through? I need somebody in my corner should mother die. I

93

know you know about psychology and psychiatry. You are probably the one to be my advocate in the family. I ended up over trusting the doctors, and I got tardive dyskinesia. I had nobody in my corner who was knowledgeable enough to help me out of it. I am sorry for criticizing you in the past. I hope to get out of that bad habit. Don't be critical of me and just accept me as I am. We have to break out of the adversarial mode in our family. I would like to hear from you to see if you would be willing to be in my corner. Think about what happened to me at Massillon State Hospital when I got no letters from my siblings for eight months, and I had nobody to turn to. I felt isolated, lonely and suffered a great deal. I need my family. I need to reestablish my relationships in the family, need to stop being critical, need to reach out simply and uncritically. I need your respect, understanding and love. I am sensitive and have different vulnerabilities in these psyche situations. I want to develop family sensitivities. I want to regain the trust of my family on a simpler level rather than intellectualize and analyze the relationships. I shouldn't be analyzing my sisters lives. It is not my place to do that. It is not my place to judge you. I want to focus on my feelings and needs. I like your mellow husband and children. The medication problems have straightened out. I am doing well in Day Treatment and therapy.

In therapy and with apologies and love,

Patrick

Timothy O'Dougherty,   7/1/96

Terence O'Dougherty and Women
        Catholics are giving in on everything over women: poverty, chastity, obedience, and the military. When we lived over on 4104 Bryant when I was four years old Terence lived with us. He was in the Chemistry Department at the time and was having some psychology problems. They were mild. Anyway, he got a few insulin shock treatments at the VA over them.   He changed noticeably right immediately. He got a job with the city rather than continue as a chemistry graduate student. He didn't like teaching. So graduate school was not the option for him. He had a 168 I.Q. on the military intelligence tests. So, intelligence was not the problem. Women were also an issue. Everybody knows about this except your family. Terry was somewhat secretive so he chose not to reveal it. He was a great man and everybody loved him. He thought that what I should do if I wanted to be a writer is to "not do experts and textbooks." I should do is to just walk out the door with little money and whatever happens to me in life happens. I did that. My father, Aquinas, liked the book, You Can't Go Home Again, by Thomas Wolfe. He thought I should use that as a tool to develop my writing experiences. Neither

94

IRISH PSYCHOLOGY/IRISH PSYCHIATRY: THE TIRESIAS COMPLEX

Terence or Aquinas wanted to be writers. They gave me the right advice though don't you think? My position is that I don't want anymore rough experiences over women after my experiences in the South. Would you? What are you giving in on over women? You can share this with whoever you chose. Nicole Brown Simpson was a Catholic.

David Noble,   7/3/96
Patrick O'Dougherty and Viktor Frankl
With my wandering and of my psychology and court experiences I ended up having Viktor Frankl like experiences in America. I have invented a new complex to explain my behavior, the Viktor Frankl complex. Like Viktor Frankl who survived a concentration camp experience, I have invented a new school of philosophy and psychology/psychiatry, the personalist intuitionist school. While I was on Bellwood Chain Gang in Atlanta in 1974, the Patty Hearst episode happened, I was beaten but not raped during this incident. I am a better person for it. I have different shortcomings, sins and faults, however, my mark as a person is charity. I tip well in restaurants, do a lot of charity work, for example, I was a big brother to a retarded youth for a year, worked with the Legion of Mary in nursing homes and hospitals for eleven years and built a major research library at Newman Center on campus gratis. I have also had many positive experiences with people. I just want to get credit for what I have done in life. My relatives were and are smart dairy farm and small town people.

God Bless, your Irish Catholic white Negro friend,

Patrick A. O'Dougherty, Ph.D.

Also, Caroline Kennedy's Secret Service agent, Officer Williams, interviewed me in the South.

Confidential!

Margaret Wright,   7/18/96

Adolescent Problems With Sexuality: "The Dark Night of the Soul"

I had different problems with sexuality when I was an adolescent. They were the dark night of my soul. I resolved them by going over them with priest confessors and with my parents. If you want to discuss them, we can. Sexuality is a "thorn in the flesh" which works against pride in our lives. It does this for me. Some of my

problems with sexuality are exiled memories which I would as soon forget. The problem that I had while I was growing up is that I came from a large family that lived in a small house. I had very little privacy. I never really had my own bedroom. This is one of the reasons that I remained single and live privately alone to this day. I am a believer in the right to privacy about my own life. I also came from a large military family that had twenty-five to thirty years active in WWII with five major campaigns. I opted for psyche consuming and analysis rather than have a repeat on this much military. One of the options my problems with sexuality led me to do in life was to lead a life of celibacy, working with the Legion of Mary and writing a book on personalism which was a philosophy favored by Dorothy Day who had an illegitimate child and perhaps an abortion. I have worked to pull women out of difficult personal situations. I haven't had any problems with sexuality since I was thirty-six years old when I opted for celibacy, the mind and will rather than the flesh. I never had many girlfriends. I never dated much. I liked books. I don't really know much about women since I have never married. I'm don't like scandals either. They bore me. I think that it is a real mistake to slip into a life of carnality. More often than not, it results in the death of the mind as well as the soul. I don't have a lot of guilt about my sexuality. However, I want to focus on the mind and soul in life-- saving my soul. Also my MMPI says that I have a very normal personality.
God Bless, to humility, to purity of intent,

Patrick A. O'Dougherty, Ph.D.

I want my sins to stay with my confessor, my psychological problems to stay with the therapists and my love to stay with my family.

Margaret, Mary Ann, and Maureen:

6/29/96
Patrick Stands up and Is Counted
In the name of Aquinas, in the name of our family, in the name of the Catholic faith, in the name of the separate way of life many of the Irish Catholics have in the world your brother, Patrick Aquinas O'Dougherty, Ph.D. stands up and is counted.

Patrick A. O'Dougherty, Ph.D.

Confidential: Between Margaret Wright and Patrick O'Dougherty

Dr. Margaret Wright, psychologist,
Dr. Jane Rozsnafszky's Professional Analysis and Opinions

IRISH PSYCHOLOGY/IRISH PSYCHIATRY: THE TIRESIAS COMPLEX

10/17/96

I brought my whole medical file over to Dr. Jane Rozsnafszky and asked her to give me a professional evaluation of it and to make some professional recommendations. She didn't write out all of her professional recommendations. However, she placed several points on me. First, she thinks the problem with Father Steve Bossi is that he was too <u>Bossy</u>. Second, she thinks I got "railroaded" in the commitment and court hearings. However, these options have worked out well for me. The situation at Newman Center resolved with both priests moving on. Third, she thinks that I have stabilized under Risperdal which is a drug. It is working out well for me. Fourth, she thinks my schizophrenia is mild. Most of the unusual thinking is due to my extensive reading. Fifth, she thinks I'm intelligent according to my tests. Sixth, she thinks my main problem is with defensiveness. This is what my personality tests show. Seventh, she thinks the personality profile of our family is critical, negative, achieving and lacking in empathy. She thinks this profile is true of my sisters, especially Margaret. Eighth, I asked her how to respond to sisters with these types of personality traits. She told me what I should do is accept them for the way that they are, don't internalize their personalities and move on in life. Dr. Werner thinks with women with these types of personalities should love their brother, Patrick. Love is most important for me, he thinks. Stacie Joncas, a counselor, also told me not to internalize my sister, Margaret's personality. Ninth, when Dr. Rozsnafsky read my file, her first reaction was that it is all critical and negative. This is the problem you have in your family. Also, you should try to come to terms with the personal validation issues for a person of your age, fifty. Tenth, she thought I have a right to privacy, and I should keep distinct what I confess to a priest and what I discuss with psychologists and psychiatrists. Eleventh, she told me she thinks that I am strong in faith and in works of charity. However, she recommends a spiritual advisor for me to work on the problem of pride. She has a spiritual advisor she told me. Twelfth, on my brother, Mike, she thinks I should just intellectually ship his wife back to Greece and just see him alone every so often. She thinks that he didn't get particularly good wives. Thirteenth, I told my mother and sisters that I love them. Dr. Werner and Dr. Rozsnafszky hope my family loves me. Fourteenth, Dr. Rozsnafszky thinks the threats issue is just legal or academic posturing on my part. I have never been violent. Also, she thinks I'm not violent and maybe less competitive and assertive than many men. Moreover, I have the trait of religiosity which she doesn't see as being pejorative. Fifteenth, Dr. Rozsnafszky thinks I'm masculine and recommends a men's spirituality group for me. She also recommends that with my unusual interests that I should get on Internet and that I shouldn't get involved in anymore power struggles. Margaret, you are a psychologist. What do you make of this evaluation and these recommendations? I don't know what to make of them. Also, I don't want a big row in our family it is going to be my fiftieth birthday this year. The counselors are impressed with me.

IRISH PSYCHOLOGY/IRISH PSYCHIATRY: THE TIRESIAS COMPLEX

Under analysis,

Patrick A. O'Dougherty, Ph.D.

Margaret Wright,   10/26/96
"Mental Illness is Not a Choice": An Apology

"Mental illness is not a choice." I didn't choose it. It happened to me when I was an adolescent. I didn't know what was happening to me at the time. Oftentimes people do not chose epilepsy or homosexuality, they are just their cross. If my problems caused you any pain or discomfort, I am apologizing for them. Out of my problems, I have made many big contributions, for example, to purity and to an intellectual synthesis in your field, psychology. I hope that I get recognition for my contributions. I just want you to be happy and to save your soul. If your husband, Frank, should die and you want to remarry it is all right with me.

Women and Paranoia
The women in our family have a problem with paranoia. They have ill founded suspicions about their brother. They don't admit that they have personal faults or make mistakes. They are critical and negative about their brother. They expect a lot out of him but then they don't give him recognition. They look to authority. They are too demanding. They should just focus on love and acceptance. Patrick is perfect the way he is.

Formidable
I don't have a problem with your husbands, Keith Nuechterlein and Frank Wright. Like your brother, they want to be formidable. Steve Allbee wants to be formidable and so do some of our Rajkowski cousins. So does Sheila O'Dougherty. However, many of our relations do not chose this option.

Mike Franey
What Mike Franey thinks I should do with my family is to move on, be there in their corner, and emphasize the positive. He doesn't think I should do all of this critical analysis about upbringing, or the divorces or give my siblings advice, or develop sibling rivalries. Critical analysis is the problem in our family not the solution.

The Father and the Mother
I decided to go the way of my parents and become a revolutionist in the Green Party.

Newman Center
Most of my big emotional and intellectual experiences have been with minorities. This was true for the Italian priest at Newman Center. With respect and love, Patrick.

IRISH PSYCHOLOGY/IRISH PSYCHIATRY: THE TIRESIAS COMPLEX

Margaret and Frank Wright,

11/25/96

"Knock the Dust off My Shoes"

I got my graduation certificate from HCMC on Friday, November 22, 1996. Joyce Forsgren, my coordinator, had a graduate school friend of hers contact my history professors to see how they evaluate me. They told her that I was brilliant. She thinks that what I should do now is to "knock the dust off my shoes" at Newman Center and at HCMC and move on to some new irons in the fire. Janis Vape, my court psychologist, dismissed my case really without prejudice on November 13, 1996.

On Advice

The advice that Joyce Forsgren gave me on giving my relatives advice was not to give them advice. Both Joyce and Dr. Rozsnafszky think that I shouldn't give Margaret and Frank Wright advice on how to raise their children or on what to do with their lives. I am going to take this advice. I won't be intrusive.

On Defensiveness

I like constructive criticism. However, the main problem that I have is defensiveness. I don't like somebody to throw cold water on me. Please keep this in mind. How do you advise people on this problem?

On Negativism

Dr. Rozsnafszky thinks it is a family problem. I am just going to try focusing on the positives. For example, I am editing out many of the negatives in the letters I've written. Bill Caldwell, a nurse, thinks it is pretty difficult to change your personality.

On the Green Revolution

I hope you like my book on the Green Revolution. I am one the major founders of the Green Revolution on campus. Did you like my data base?

Happy Thanksgiving and I hope that I get to talk to you soon,
Love,

Patrick

Margaret,

A Happy Time

What I wanted in life was to get a Ph.D. and to be out of the slot and a writer. I got that. In high school my dream was to be a theoretician or revolutionist or activist in an international political or historical movement like the Green Party. I have solidified my position and am embarking on that goal. I also wanted to help

99

define the Irish white Negro cause as an intellectual. I am achieving that. It is a happy time for me.

Friends

I don't have many close friends. However, I do have close friends from all of my stages of life: grade school, my high schools, St. John's, the University of Minnesota, Graduate School, Harvard, political friends, friends from my psychology experiences, writer friends, and clergy.

The Empathy/Sympathy Issue

I feel that our family has many strengths. They are religious, intelligent, good looking, hard working, international and psychologically minded. However, I don't feel that the women feel particularly secure with a single celibate male like me. Do you have a problem with this? Oftentimes it is the single people in a family who weather out the financial storms well and who achieve well because they have their freedom. This maybe true for me. I feel that I am the third spoke missing on a three spoke wheel with the married couples in our family. I also feel that the siblings are not particularly empathic with me. I will give you an example. I was in Massillon State Hospital in Ohio for eight months and I never received a card or letter from any of my siblings. Would you like to be in a hospital for eights months and not receive any mail from your siblings? I had tardive dyskinesia for more than eleven years and I received no recognition for my pain. However, I am a masculine guy who likes to meet his own problems head on. The deal at Newman Center is the military man picked a fight with me subtly and nobody in our family took my side or put a good word in for me. Would you like that to happen to you? I am hard on myself and my family is hard on me.

The Divorces

The advice I've been given on the family divorces is "judge not lest ye be judged."

On Giving Advice

The advice I've been given on giving my siblings advice about their spouses or children is "don't give advice."

Liked the Priests

I never was orient towards being a jock or towards the military, however, I always liked the priests and nuns. I like the spiritual disciplines.

The Eldest and the Burden of Status

Both of my parents and my siblings were tough on the eldest for the burden of status. I want the family to ease up.

100

IRISH PSYCHOLOGY/IRISH PSYCHIATRY: THE TIRESIAS COMPLEX

The Psychologists/Psychiatrists
The therapists aren't for this critical, negative, faultfinding, little praise family I've got. My therapists are accepting, positive, nurturing and validating for me.

Keith and Frank
Your husbands impress me. I like them. However, I'm different from them.

A Positive Influence on the Family
I did the businesses in our family. I championed the spiritual dimension. I am a teacher, activist and writer. I'm single. These are all positives.

On Margaret
Margaret is good looking, intelligent, accomplished, a good mother, and has a fine spiritual dimension.

Dr. Frank Wright,   5/1/97
Second Chances
Almost none of my father's family, the O'Dougherty family, were for the second marriage of Margaret O'Dougherty and Frank Wright. I went along with it. The Pope says that divorced and remarried Catholics should abstain from sex and should not receive the sacraments. Margaret goes to communion. This is sacrilegious. The Catholic Church is a sacramental church which means a mediated relationship with God. Many Protestants have an unmediated, personal relationship with God. If you were a physician in private practice, would you like it if an employee of yours embezzled money from you and just kept this sin between him and God? Many people rationalize theft. The Church doesn't want people to steal or murder and just kept a personal relationship with God about it. This is why we have the sacraments. The O'Dougherty family were always quite encouraging with me. In contrast, the Coynes were not as much. Mabel never called me, or sent me a card, or gave me a gift. Phyl was quite aloof. Donna and my mother were better. I like to stand up for the O'Dougherty's because they encouraged me. I am not trying to make and affront to you. I wish you and Margaret well in your marriage. If you want me to have little involvement in it that is okay with me. I think you are an intelligent, decent professional man. Two thirds of second marriages end in divorce. I think this could be very hard on children. The main opposition to the Catholics in this country are the Presbyterians, the Baptists and the Unitarians. You married a Scotch Presbyterian. That was your cross in life. My cross is schizophrenia. As I mentioned before, Keith Nuechterlein sent me a very encouraging Christmas card.

Frank Wright, Physician

IRISH PSYCHOLOGY/IRISH PSYCHIATRY: THE TIRESIAS COMPLEX

I am a pro-life Catholic. Pretty soon every GP is going to be able to issue prescriptions for the abortion pill. What do you think about that? The Northwest Territories in Australia have legalized voluntary euthanasia. The liberals are going to legalize it in this country. There is going to be a medicide clinic in every minority, gay and elderly community. I am not going to be for euthanasia or medicide clinics. I have had some difficult problems with schizophrenia and tardive dyskinesia. I had drug related thoughts about suicide. Why don't you see the film <u>Shine</u> or the film <u>Surviving Picasso</u>? The leading character in the film <u>Shine</u> had schizophrenia. Mark McGee and I think Picasso was an intense schizophrenic also. If a person with schizophrenia like me, or like the lead actor in <u>Shine</u>, or like Picasso, or gay or a minority with schizophrenia came to you as a Catholic physician with some very difficult problems would you refer the person to the neighborhood medicide clinic? Peg doesn't want me to visit because she thinks I might say something inappropriate or be too controversial. She isn't schizophrenic. Neither are you. Would you like this? Do you think somebody wants to have schizophrenia or epilepsy? I think the Wrights should be compassionate and loyal to me. I am to them. Life is going quite well for me. I have my publishing company, a new job, and many friends and groups that I'm involved in. I am getting professional recognition. I'm getting a second chance. I'm glad we have a physician in our family.

With prayers and love,

Patrick
cc. Margaret Wright

CLOSURE
MY MAIN MEMORIES OF MY FAMILY WHEN I WAS GROWING UP ARE QUITE FOND MEMORIES. THE RELATIONSHIPS IN MY FAMILY WERE AND ARE ANOINTED. MY WHOLE FAMILY IS CALLED TO HOLINESS. THE O'DOUGHERTY FAMILY IS A SPIRIT FILLED FAMILY.

LETTERS TO CLERGY

May 9, 1991

Patrick O'Dougherty
Riverside Plaza M3410
1615 South Fourth St.
Minneapolis, Minnesota 55454

Father Virgil O'Neill
St. John's University

IRISH PSYCHOLOGY/IRISH PSYCHIATRY: THE TIRESIAS COMPLEX

Collegeville, Minnesota 56321

Father Virgil,
    It was excellent seeing you again on theology day at St. John's. A friend of mine made the reservation for me. I have been a practicing Benedictine Oblate for eight years. I expressed an interest to you about making a private profession of vows for one year up at St. John's. My year of waiting will be up in the fall and I do it then. Father Michael Skluzacek was one the presiders at my final Oblation. I met the Abbot, Father Theisen, at this ceremony. I wonder if you could put me in touch with someone at St. John's Abbey who could help me take these vows?

    Today is my sister, Maureen's, birthday. She is becoming an anthropologist out in Manhattan. I sent her a copy of my book, <u>Walden III: A Catholic America</u> for her birthday present. My mother and Dick Kast her new husband went to the Florida Keys for a trip. I went to see Shakespeare's play, <u>Julius Caesar</u>, two weeks ago. A friend of mine, Thomas Linstroth played Julius. Did you ever know any of the younger Rajkowskis, Frank and Mike, while they were at St. John's? Both of them became lawyers.
My mother and Charlotte say hi.

    I would appreciate it if you would write a response to this letter when you have the time.

    To the last Benedictine in America, sincerely yours,

    Patrick A. O'Dougherty

7/17/91

Patrick A. O'Dougherty
Riverside Plaza M3410
1615 South Fourth Street
Minneapolis, Minnesota 55454

Father Magnus Wenninger, O.S.B.
St. John's Abbey
Collegeville, Minnesota 56321

Father Magnus,

IRISH PSYCHOLOGY/IRISH PSYCHIATRY: THE TIRESIAS COMPLEX

Thank you for your letter of 7/14/91. You wondered in your letter why I want to make a private profession of vows. The reason is that I am a writer. It would help me out if I could refer to myself as a brother on <u>some</u> of my books. Can I do this as an Oblate? I am going to be publishing a trilogy in American history in the fall or next year. The books that I have published so far are: <u>An Existential Approach to American History</u> and <u>Walden III: A Catholic America</u>. I am soon working on a book on the Progressive Period in American history.

Also, I want to make a deeper commitment to Christ. I try to go to mass daily. I worked in the library up at Saint John's with Father Urban when I was an undergraduate; I want my books placed in the library. Did you ever know my dad, Aquinas O'Dougherty, when he was up at St. John's? He was one of the editors of the <u>Record</u>. It is possible that I could make a directed retreat up at St. John's in the fall. My father is dead.

Sincerely,

Patrick A. O'Dougherty

P.S. I think that you are a pretty unusual man. I cannot get this letter mailed right away because I am having knee surgery.

7/26/91

Patrick A. O'Dougherty
Riverside Plaza M3410
1615 South Fourth Street
Minneapolis, Minnesota 55454

Father Magnus Wenninger, O.S.B.
St. John's Abbey
Collegeville, Minnesota 56321

Father Magnus,
Thank you for your letter of 7/24/91. Just out of curiosity, haven't you heard of Secular Institutes. They are a way that an individual can make a private profession of vow and live in society. I think that Pope Pius XII gave him his approval. There are several them in Europe and several of them in America. Does

IRISH PSYCHOLOGY/IRISH PSYCHIATRY: THE TIRESIAS COMPLEX

St. John's Abbey sponsor one of these? Does the Benedictine order consider them to be worthwhile? Aren't these Institutes defined in Canon Law?

Sincerely,

Patrick A. O'Dougherty

7/14/91

Patrick A. O'Dougherty
Riverside Plaza M3410
1615 South Fourth Street
Minneapolis, Minnesota 55454

Magnus J. Wenninger, O.S.B.
St. John's University
Collegeville, Minnesota

Magnus,

I am writing to request the brochure: "You and God, Opportunities for Growth," issued by The Spiritual Life Program of Saint John's Abbey, Father Don Tauscher, director. I read about this brochure in an Open Letter To Oblates that you wrote in the St. John's Abbey Quarterly.

I would also like to visit St. John's University in late August or early September to renew my Oblation. Furthermore, I want to make a private profession of vows of poverty, chastity, and obedience for one year. I want to live in society with these vows. I made a final profession as an Oblate of St. Benedict on June 3, l983, with a group of Oblates from the Cathedral of St. Paul. I have been attending monthly meetings since then. I have waited a year after deciding to make this private profession. I would like your aid in this decision.

With Benedict,

IRISH PSYCHOLOGY/IRISH PSYCHIATRY: THE TIRESIAS COMPLEX

Patrick A. O'Dougherty

Patrick O'Dougherty   6/7/92
Riverside Plaza M3410
1615 South 4th. St.
Minneapolis, MN 55454

Brother Gregory Conant
252 Still River Road
Post Office Box 67
Still River (Harvard)
Massachusetts 01467

Brother Gregory,

I went to a performance of Bernstein's Mass. Have you ever heard it? Do you think that it is blasphemous? He has Persian and Oriental music in it as well as pop and rock music. There were about 150 performers in this Mass. I like Shubert's Mass better.

I also went to a performance of Carmina Burana. The music of this performance based itself on 24 poems found in the monastery of Benediktbeuern. The title translated means "Songs of Beuren." Carl Orff built his music around these poems or songs. There were two hundred voices alone in the performance. The work is in Latin and German. I was taken back by this music. Do you like classical music?

I understand the Jesuits as well as the Benedictines trained you. What sort of curriculum did they have at Siena? What did you major in there? I will call up your priest friend sometime this summer. I guess that Ken has a bad leg. My students had a difficult time with division of polynomials. They have two new priests at Newman Center so I do not want to tell them how to run the center right off the bat. I have been to Boston three times. You are welcome to meet my friends and associates. If you wish to give a talk here, I will try to arrange it. I was born in St. Cloud. Do you think that Freud would work well with an analysis of the Founding Fathers? The American Revolution was a revolution within--a psychological approach.

I guess Ann Conant is going back to college. I live on the 34th. floor of Riverside Plaza. I think that you should have majored in chemistry. Your mind is an element. Are you coming to Minneapolis soon? Hope to hear from you soon. I send you my manuscript when it is ready.

IRISH PSYCHOLOGY/IRISH PSYCHIATRY: THE TIRESIAS COMPLEX

You are in my prayers,

Patrick O'Dougherty

Brother Gregory, 11/4/93

    I sent your letter and essay up to Sister Mary Anthony Wagner. I hope that she publishes it. Your intellect impresses me; and, I have saved most of your letters. My parents thought the world of the Conants.

    I am trying to build a Hellenic Synthesis in America. That is the name of my company and institute. I have published four books and I hope to make a contribution like Orestes Brownson and Soren Kierkegaard did. Brownson wrote forty to fifty small books and essays and Kierkegaard developed a little dialogue within the Danish Christian community. I hope that my little synthesis becomes a black classic. It is a foil to the nonwhite contribution in America.

Write me soon, a Conant lover,

Patrick A. O'Dougherty, Ph.D.
American Intellectual Historian

Sister Mary Anthony,   11/10/93

    I read your book, The Sacred World of the Christian, and I liked it. However, I think you could strengthen it in the area of interpretation. How would you differentiate Christian approaches to anthropology, specifically American Indian rituals, from Claude Levi-Strauss' structural anthropology? I found that it was not self-conscious enough about methodology. I'm a Christian existential historian. This means doing construct validity with existential themes.

God Bless,

Brother Patrick O'Dougherty

8/11/94

Patrick A. O'Dougherty, Ph.D.
Riverside Plaza M3410

IRISH PSYCHOLOGY/IRISH PSYCHIATRY: THE TIRESIAS COMPLEX

1615 South Fourth Street
Minneapolis, Minnesota 55454

Catholic Bulletin
Letters to the Editor

Editor:

I want to make a small criticism of the <u>Catechism of the Catholic Church</u>. I have been going through the Index of Citations at the back of the Catechism; and, I have found very few references to anything American or Latin American. Under the topic Congregations there is a Letter to the Archbishop of Boston dated August 8, 1945. There is a reference to <u>The Apostolic Fathers</u> in the Abbreviations. However, the word "Liberty" is not in the Subject Index. The Latin Americans have one saint mentioned, St. Rose of Lima. There have been Catholics in the New World for five hundred years. I think the Americans should be given much better recognition by the Church. For example, I think they should hold a Particular Council or Synod in the New World. As an aside, I think in the interest of diversity American Catholics should work to have clergy on the staff of the University of Minnesota. They need an American Catholic renaissance.

Patrick O'Dougherty, Ph.D.

Patrick O'Dougherty, Ph.D.   10/8/94

Most Reverend Lawrence Welsh
Bishop of Saint Paul and Minneapolis
226 Summit Avenue
Saint Paul, Minnesota 55102-2197

Excellency:

Thanks for your phone call. I am trying to get the first priest on the faculty of the University of Minnesota. The priests were some of the first people in this state besides the Indians. I happen to like the priests. I have published five books. I want to publish as many books as Orestes Brownson did. I like the great martyrs and the great intellectuals.

With the Love of Christ,

IRISH PSYCHOLOGY/IRISH PSYCHIATRY: THE TIRESIAS COMPLEX

Patrick O'Dougherty, Ph.D., historian

Archbishop John Roach,   12/18/94

Remember the Hindenburg Blimp

Hydrogen is the most common and the most efficient element in the universe. Every country has it--Latvia, Angola, Samoa, etc. It is the fuel of the stars. **The reason that we don't use it in automobiles is because of the Hindenburg Blimp accident. I think that we should change our economy over to hydrogen fuel as rapidly as possible.** Why? This state had the world's highest concentration of high grade iron ore on the Mesabi range. It went in thirty years from 1915-1945. We cannot pass a law to get it back. It was useful for surgical tools in the Third World. My uncle, Monsignor Loyola O'Dougherty, thinks that the same deed will happen to the copper in Arizona where he lives. This is going to happen to our oil. If we don't switch over to hydrogen fuel, many poor countries will become slaves to the Arab oil interests--maybe just through banking interest. The Catholic lives for the Common Good. There are always military options to solve economic or population problems or there is the option of colonialism. China has very few oil resources. If we don't switch over to hydrogen fuel, billions of people will suffer. There is little proof for the idea there is an energy crisis. It is a lie. **There is very little democracy in nature.** I am not exaggerating here at all. If you want a reference on this, call Dr. Mike Franey at the University of Minnesota Physics Department. Tom Linstroth, a St. Thomas Academy student, will tell you the same.

Merry Christmas,

Dr. Patrick O'Dougherty, Ph.D.
The Holy Cities in the New World are Mpls. and St. Paul, right?

Archbishop John Roach,   1/13/95
The Newman Papers
There is an interesting book by Fionnbarra O'Dochartaigh, ulster's white negroes: from civil rights to insurrection. He is the co-founder of the Northern Ireland Civil Rights Association. He parallels the Catholic experience in Northern Ireland to the struggles of black people throughout the world. He says "We viewed ourselves as Ulster's white Negroes--a repressed and forgotten dispossessed tribe captured within a bigoted, partitionist statelet that no Irish elector had cast a vote to create..." He tries to be **scientific** in his presentation, and he relates his data to "the massacre in 1972 of thirteen unarmed demonstrators on Bloody Sunday."(cover)

IRISH PSYCHOLOGY/IRISH PSYCHIATRY: THE TIRESIAS COMPLEX

My uncle, John O'Dougherty, is not at all impressed with Daniel Patrick Moynihan's mind. The Kennedy children are well through after the assassinations of their fathers my dad thought. The Irish are related to kings. Pope Paul VI called for a few good martyrs.

Father Michael Tegeder is for changing over to hydrogen fuel, a Nobel Prize class for women, sending Europe's royalist heritage to America, the Catholic causes in the British Isles, a priest on the faculty of the University of Minnesota, a Holy City in the New World, and a Catholic rationalization of the American experiment. I talked to him about this. He was a priest in England for a while.

I heard Msgr. Schuller give a sermon on <u>Everyman</u>. Everyman leaves faith and hope behind and brings his gifts of charity to heaven. If you are going to write some memoirs, try the theme of love. St. John did.

There is a premium on intellectual contributions in the world.

Patrick O'Dougherty, Intellectual Historian, Ph.D.

Archbishop John Roach,   1/12/95
The Newman Papers
I want to let the situation at Newman ride for a while. The building is empty most of the week. There are just a few students from the neighborhood who go to the coffee house. Most of them are probably not Catholics. My aunt, Charlotte O'Dougherty, thinks that $196,000 is way too much to underwrite the Newman Center. We have three law firms is our family maybe you should get to know one of them? My father, Aquinas O'Dougherty, a University of St. John's graduate, was the business attorney for Regina High School. After he died, the school went under.

Do you remember Tom Linstroth, Mike Franey and George Norris from St. Thomas Academy? They all wish you well, especially Tom Linstroth. Do they teach principles of genetic warfare at St. Thomas Academy?

The way that you can test a priest is by love. You could write about love and the impact of the Benedictine way of life on your episcopacy. The Bishop who impressed me is Bishop Jean Baptiste Lamy of Sante Fe. Georgia O'Keefe lived there. Los Alamos is in the vicinity. St. John's is there. They teach the classics at St. John's.

My roommate from St. John's, John Herlick, thinks that we should changeover to hydrogen fuel in our automobiles. It would be a trillion dollar deal. They don't have many fossil fuels in Brazil or East Germany. We should conserve them.

110

IRISH PSYCHOLOGY/IRISH PSYCHIATRY: THE TIRESIAS COMPLEX

Have you read the book by Finnabar O'Doherty on the Northern Irish as "white Negroes." He helped found the Northern Irish Civil Rights Association. He wants to internationalize this movement.

Patrick A. O'Dougherty, Ph.D.

Sister Mary Anthony: 1/14/95

"Act Locally, Think Globally"
St. Benedict's Convent St. Joseph, Minnesota/World Spiritual Dominion

I hope to see a woman Doctor of the Church come out of the Benedictine Community at St. Benedict's College. I think the Benedictine women have a lot of good material up there. I'm not in your field so I can't rate your material critically.

A New Catholic Philosophy
I am an activist/writer with the International Alliance for Sustainable Agriculture. I have built an internationally recognized library with their group in the Dorothy Day room at the Newman Center at the University of Minnesota. The book I gave you presents a new philosophy in Catholicism--personal intuitionism. Women are noted for being good at languages. I chose mathematics as the language for this philosophy. It is a counter to the Communist Manifesto. It is a personal Declaration of "Interdependence." I want to have our Oblates of St. Benedict group meet in the Dorothy Day room at Newman Center some Saturday. It is at 1701 University Avenue S.E., Minneapolis. As an aside, what do you think of Caroline Kennedy's effort to move these America men beyond superman?

Back the Green, the Grey and the Purple Evolution/Viconian Revolution,

Patrick A. O'Dougherty, Ph.D., The Hydrogen Conversion Project and the Alternatives to Petrochemicals Project, Towards a Global Economy Based on Self-gift rather than Self-interest.
Do you have these Fax numbers?
Fax: Pope John Paul II, The Apostolic Palace, (011)39669.8850.88
Fax: Nobel Prize Foundation, (46) 86603847

Archbishop John Roach

IRISH PSYCHOLOGY/IRISH PSYCHIATRY: THE TIRESIAS COMPLEX

The Chancery
226 Summit Avenue
St. Paul, Minnesota 55102

3/3/95

Your Excellency:

I am formally requesting a legal hearing in the Priest's Senate or in the Court of Canon Law about the situation at Newman Center. I want to make a legal record of my position and my ideas. They are asking for a lot of money from the archdiocese, $199,000. I want to give them a chance to brace their positions. I have had many legal cases. This will be fun. Here are brief presentations of my ideas and positions. I am just trying to do the honorable. However, I know that law suits can cost more than $199,000. I am going to write the heavyweights in the Church about my case. We will see what happens.

An Admirer and Critic of Yours,

Patrick A. O'Dougherty, Ph.D.

Patrick A. O'Dougherty, Ph.D.
Riverside Plaza M3410
1615 South Fourth Street
Minneapolis, Minnesota 55454

Archbishop John Roach
Bishop Lawrence Welsh
Bishop Harold Flynn

3/5/95

Excellencies,

Major Embezzlement and Theft at Newman: Inside Job

I am a historian/activist with the International Alliance for Sustainable Agriculture which is in the basement of Newman Center. I have a pretty good idea about what is going on at Newman Center. There was a major embezzlement and computer theft that occurred recently at Newman Center and at the International Alliance for Sustainable Agriculture. Joseph Barisonzi the head of IASA at the time told me that a man named Williams wrote checks on the Newman Center account and the IASA accounts and used his keys to steal a computer and computer equipment. I heard that Tom Conry's office was broken into. The theft of funds and the computer property amounted to a major felony. The police heard of this. It was an inside job. Is there a coverup? Is the theft covered by insurance? The individual

112

implicated fled to avoid prosecution. Is he being prosecuted? There is no honor among thieves. He will hurt somebody else if there isn't a follow-up. Were funds of the archdiocese stolen? Has there been an internal audit? IASA is thinking about leaving Newman Center. On whose watch did this happen? I think that they should come clean. This is all a matter of record. It occurred several months ago. Joseph Barisonzi left as the head of IASA. Now Newman Center is asking the archdiocese for more money. What am I supposed to think? Newman is way in the red. What happened?

Patrick A. O'Dougherty, Ph.D.
Joseph Barisonzi told me that just one of the checks Williams wrote on IASA was for $700. He told me that Newman Center lost much more. This is how they identified Williams as the computer thief.
Patrick A. O'Dougherty, Ph.D.

Archbishop John Roach
Bishop Lawrence Welsh
Bishop Harold Flynn
3/8/95
Excellencies:
Brechtian Liturgies at Newman Center
Tom Conry is the liturgist at Newman Center. He told us that he is having Brechtian liturgies at Newman Center. He builds skits and dialogues around various Biblical themes with a Brechtian structure. Bertolt Brecht was <u>not</u> a liberal. He was a major **Marxist socialist/communist dramatist**.   His plays have four main characteristics: "I, the epic element by which the plays emphasize pointed presentations of social conflicts and treat dramatic protagonists as mere exponents; 2; the didactic element, by which the plays attempt to educate the audience concerning its social position; 3, music to provide a greater emotional impact; and 4, adaptation of stories and plays by other writers," in this case Biblical writers. (Source: Collier's Encyclopedia, 1959, Vol. 4., 48). See Tom Conry's enclosure, "Looking at Liturgy: Aristotelian vs. Brechtian Lenses." Is he setting up a dialogue or dialectic?
A Deconstructed Brechtian Liturgy at Newman Center
Deconstruction is "a philosophical movement and theory of literary criticism that questions traditional assumptions about certainty, identity, and truth, asserts that words [refer only to words,] and attempts to demonstrate how statements about any text subvert their own meanings." Source: American Heritage Dictionary, Third edition, 1992. Also Random House Dictionary. For example, Dr. Mike Franey, physicist, suggests the Constitution or the Bill of Rights, i.e, the right to privacy, deconstructs. Do the ideas of Aristotelian vs. Brechtian "a realistic skepticism" and

113

IRISH PSYCHOLOGY/IRISH PSYCHIATRY: THE TIRESIAS COMPLEX

the "assembly is 'colony'; a "client class'" found on Tom Conry's flyer illustrate this idea? The answer is yes. Church dogmas or liturgies can deconstruct. Brechtian liturgies at Newman have deconstructed my ideas of assembly, sacrament, and the clergy/laity relationship. In Rome, they had "bread and circuses" for the masses. Is this what they are going to have at Newman Center? Patrick A. O'Dougherty, Ph.D., Intellectual Historian.

Archbishop John Roach: 8/9/95

Carol, your secretary, suggested I write up my ideas and send them to you. Father Freeman has a concern that the next pope will be the first modernist pope. He told me that if the modernist cardinals get one third of the vote in the next papal election plus one the modernists will force the election of the first modernist pope. Do you favor a modernist pope? My concern is the next pope is will be the first postmodern pope?

Father Francis Fleming, retired pastor, St. Olaf's Church
Father Fleming read <u>Ms.</u> magazine and told us in a sermon that he was for about 70% of the issues in it. How would you address the issues in this magazine? I am for 80% of their causes.

The Austrian Studies Department
Pieter Judson, Swarthmore College, is giving a lecture in the Beyond Vienna 1900, Austrian Studies, Robert A. Kann Memorial Lecture Program, called "Dangerous Cultures: Rethinking the Liberal Legacy to Vienna 1900." Is America's liberal legacy a "dangerous culture?"

The Irish Institute at St. Thomas College
Why is the Irish Institute leaving St. Thomas College?
I told Brother Gregory, a history major at Harvard, to address the Trotskyite movement. He is looking into it.
Brother Gregory is a "prophet armed" intellectually.

Patrick A. O'Dougherty, Ph.D., mock Nobel laureate
Catholics love theology, philosophy, and intellectual history.

Archbishop John Roach
Bishop Lawrence Welsh
Bishop Harold Flynn

Excellencies: 3/18/95
The Strength of the Irish People

IRISH PSYCHOLOGY/IRISH PSYCHIATRY: THE TIRESIAS COMPLEX

My father, Aquinas O'Dougherty, thought a basic strength of the Irish people is that they have a cause or causes. He also thought that the basic problem that most people and many peoples have is that they don't have a cause or causes.

Project Nobel Prizes

The congregation at Newman Center should start Project Nobel Prizes: sweep all the mock Nobel Prize classes and create a new class for women's contributions. This is a noble cause.

Evaluating Archbishop John Roach's Tenure

I have not studied Archbishop John Roach's life and writings. Moreover, it is difficult to make to the position of archbishop. He led the National Conference of Catholic Bishops. I have read some of his writings, "Reviving the Common Good," and I think I read an essay he wrote on, "Love and Sexuality." I also have the Pastoral Letter, "Economic Justice for All: Pastoral Letter on Catholic Social Teaching and U.S. the Economy," which he helped produce. I liked Renew. I always liked Archbishop John Roach. The Catholics produced many of the classics in Europe. They want one in St. Paul/Mpls. The archbishop was the head St. Thomas Military Academy in his younger years. Furthermore, I heard that the archbishop likes cards. Does he play, "blackjack?" How is that for a Catholic classic?

Do or die, the solitary eagle seeks the flame,

Patrick A. O'Dougherty, Ph.D.

Patrick A. O'Dougherty, Ph.D.

Archbishop John Roach
Bishop Lawrence Welsh
Bishop Harold Flynn

3/23/95

Excellencies:

Recent Staff Defections at Newman

Terry Knaffla
Daniel Fleming
Jacqui Landry
Jim Fournier
Julie Bates

115

IRISH PSYCHOLOGY/IRISH PSYCHIATRY: THE TIRESIAS COMPLEX

Marnie Kelly
Terry Gips
Susan Heidenreich
    Why is it that anyone with any ability is leaving Newman Center? Is it a Christ versus Caesar problem? Is it a salary problem? Is this a normal turnover? Do they just want yes persons?

    The main purpose of Newman Center is to brace the atheism/agnosticism and secularism on campus. We want a scientist priest on the faculty of the University of Minnesota. Catholics have paid a price in America and Minnesota. The clergy are often the finest. We want a clergyman on the faculty. Father Michael Tegeder told me a priest founded the beekeeping field of studies on the St. Paul Campus.

Patrick A. O'Dougherty, Ph.D.

Archbishop John Roach
Bishop Lawrence Welsh
Bishop Harry Flynn
5/24/95

Excellencies:

Beyond Catholic Marxism at Newman

    Father Steve Bossi's tenure at Newman has been Catholic Marxism. I feel that he is in a quagmire. I want to expand the picture at Newman Center to the whole of Catholicism not just some left-wingers at Newman Center. For example, I heard that Princeton University had Opus Dei in their graduate school. I asked Terry Dosh to invite them into Newman Center to inspire the young people. He almost had a seizure. I don't know much about Opus Dei. What is wrong with Opus Dei? I asked Father Steve to invite some Catholic Chicago school economists to talk at Newman. He told me that it would be "divisive." I asked Glen Smoot about inviting somebody from the Wanderer or the Remnant to talk at Newman. He blanched. Let's have Fidelity Magazine, the Homiletic and Pastoral Review, The Blue Army, Apropos, The Legion of Mary, The Mindszenty Report, The Catalyst, Culture Wars and the ACLA Report at Newman. Let's look at German Catholicism, and Irish Catholicism and Samoan Catholicism. Archbishop John Roach, I think, started a group of Brothers to work against abortion. Let's get them in at Newman.

Charity Not "Crossing the Rubicon" at Newman

    Dr. Mark McGee thinks that they aren't treating me very Christian at Newman because I raised the money issue. I want to experience charity and kindness not "crossing the Rubicon" at Newman Center.

116

IRISH PSYCHOLOGY/IRISH PSYCHIATRY: THE TIRESIAS COMPLEX

My library that I built gratis at the International Alliance for Sustainable Agriculture is a life and death issue at Newman. Let's look to its future.
From The Project to Refound World Culture, Maximum Intellectual Risk All Issues, Patrick A. O'Dougherty, Ph.D.

Archbishop John Roach
Bishop Lawrence Welsh
Bishop Harry Flynn

Excellencies:

5/30/95
Catholic Marxism to Dionysus Veiled
At Newman Center, they have Catholic Marxism. Moreover, in their liturgies they have burning caldrons, Greek dialogues, a quasi South Dakota Passion Play, the voice of God from a cherry picker, and the song cycles in the Mass are like a hootenanny. Is the consecration a footnote? Also, they have the grey reality.
The Writer's Craft
Patrick is every type of a person. I have specialized in counseling young people, and in teaching at all levels, and I have taught writing. I want to be nondirective, brace all intellectual positions, and let the students come up with their own material. I don't claim any of them. I have done quite well with this approach.

A Separate Way of Life
The Church is about a separate way of life in the world. This is what the South is about too. For example, there is a famous Southern writers book called, I'll Take My Stand. It is declarations for the Southern way of life. I think Allen Tate, a famous Regents Professor at the University of Minnesota, features in it.

America's Strengths and Weakness
America's strength is in technology, inventions, and applied science. Compared to Europe, America has very few intellectuals and artists. I used to think Americans were good at government. However, Plato thought a republic would last about 200 years. After that the bureaucracy would become so large as to make the republic unworkable. The 200 years are up.

God Bless, With Spiritual, Intellectual, and Psychological, Courage,

Patrick A. O'Dougherty, Ph.D., mock Nobel laureate

Sister Mary Anthony   7/13/95
St. Benedict's Convent

IRISH PSYCHOLOGY/IRISH PSYCHIATRY: THE TIRESIAS COMPLEX

St. Joseph, Minnesota 56374

Sister Mary Anthony:

I do research in intellectual history and the history of science with David Noble and Mike Franey at the University of Minnesota. Bishop Lawrence Welsh has shown some interest in me. Could you tell me how I could look up his publications? I have Margaret Mary Reher's, <u>Catholic Intellectual Life in America</u>. Cardinal Dennis Dougherty and Cardinal William O'Connell made the cut.

Patrick A. O'Dougherty, Ph.D.

Archbishop John Roach
Bishop Lawrence Welsh
Bishop Harry Flynn

6/2/95

Excellencies:

Maria Carolina Torres

We had a Catholic foreign exchange student, named Maria Carolina Torres, from Recife, Brazil, live at our home for one year during the 1960s and attend Regina High School. While we had protests in the USA during the Vietnam War era, the protests in Brazil resulted in many people disappearing from her town. How would you address this problem? It is the theme from a movie I saw in Paris called <u>Kiss of the Spider Woman</u>. Other themes from the movie are "intellectual courage" and the masculinity issue. It is about a Marxist or leftist and a homosexual in a prison in Brazil. I wouldn't recommend this movie to a bishop. However, my mother, told me that a Cardinal in Brazil is trying to recover the disappearing generation in Brazil and in Latin America. Carol moved to Paris to live.

Patrick A. O'Dougherty, Ph.D.
Dear Bishops: Roach, Welsh and Flynn 6/15/95

Newman Center: A Coup de Grace?

Please tell the staff at Newman Center that a coup de grace or a coup de main is inappropriate for a Catholic Center. The young people want a coup de foudre or a coup de maitre.

Nicole Brown Simpson

Nicole Brown Simpson was a Catholic school girl. I heard that she married in a Catholic Church. Who approved this? Brown is an Irish name. Nicole is a European name.

Single People

Single people more than 18 are in the majority now in the United States.

Patricia Coyne

118

IRISH PSYCHOLOGY/IRISH PSYCHIATRY: THE TIRESIAS COMPLEX

My mother is writing her autobiography. I recommended the themes from St. Benedict's Rule to structure her story around. There are many of them.

Lenin's Marxist Conquest of America: Clinton and Abortion

One of the first acts Lenin did when he assumed office was to legalize abortion. He was one of the first of the European leaders to do this. Now we have this Marxists/Leninist holocaust in America. It is part of our Vietnam legacy and Clinton's legacy.

The American Legal System

The Vatican condemned the American legal system several years ago saying that it is "'corrupt and morally bankrupt.'" (Source John O'Dougherty) The legal system in this country has turned into a military defeat. It is an internal Vietnam.

"Anti-Intellectualism in American Life"

There is a famous work in intellectual history by Richard Hofstadter with this title. Why don't Americans like intellectuals that well? They like the practical sciences and pragmatism. They are good at inventions. They don't have a real problem with them in Europe. For example, many of the Protestant religious sects are unsophisticated intellectually and theologically compared to the Catholics and the Lutherans.

The Chump Issue and Catholicism

Catholics get very little in the public schools, in the media, in the intellectual culture of America. How can 50 to 75 million Catholics possibly bail out "failed government programs costing trillions?"(John O'Dougherty) I think that what Catholics get in America is "soldier's pay."(Quote: William Faulkner) Christ was not a chump. I'm not either. Patrick A. O'Dougherty, Ph.D.

Archbishop John Roach
Bishop Lawrence Welsh
Bishop Harry Flynn   **Happy Father's Day**
Dear Bishops: 6/16/95

A Space Center at St. John's University

I'm for a space center at St. John's University to make Catholic schools a leader in scientific education.

American Conversion

I'm for a partial conversion of the ICBM's in the Dakota's and out West to a Space Center in the North for the peaceful, scientific and spiritual conquest of space--a Nobel Peace Prize.

Married Priests

I am never going to be for married priests. My father had me pay too big a price. I'm not marriage directed. Terry Dosh was a teacher up at St. John's who had interest in me. He is for married priests. However, I don't know about his newsletter "Bread Rising" at all. It is Catholic thinking fads.

A Catholic "Counter Revolution"

IRISH PSYCHOLOGY/IRISH PSYCHIATRY: THE TIRESIAS COMPLEX

I like Eugene Michael Jones and John Cardinal Krol. They are tremendous Catholics. I like Pope John Paul II. I like Joseph Ratzinger.

"Book learned and textbook learned" Catholics (Quote Terence O'Dougherty)

My uncle, Loyola O'Dougherty, is a famous priest in this country. For some reason his nephew wanted to be a writer and write a classic, like a "black classic." He does not have a problem with me. I like Augustine, Aquinas and St. Louis Marie De Montfort. However, I taught writing, and you can't just learn to write from books and textbooks. You just have to go out and experience life's situations. I did just that. I don't want to be just a "book learned or textbook learned" Catholic. I want to break at least 1000 paths in my fields--intellectual history and the history of science. Let the Bishops separate the "wheat from the chaff."

Bishop Larry Welsh, Thanks for the call. I told my physicist friend, Mike Franey, about you. My position is we have interest in you.
Patrick Aquinas O'Dougherty, Ph.D. mock Nobel laureate

Bishop Lawrence Welsh: 6/23/95

The PostColonial Interpretation and Catholicism

Basically, the Irish and the Polish peoples in Europe and America were the suppressed, subjugated, and colonized peoples. The literature that has come out throughout the world since the presidency of JFK and the Papacy of John Paul II is postcolonial literature. For example, the literature of the Polish woman today is postcolonial literature. Pope John Paul II is a postcolonial pope. One interpretation of his life and writings is that they are a postcolonial theology, philosophy and prayers.

American Catholicism: A Question Mark?

My uncle, Msgr. Loyola O'Dougherty, thinks American Catholicism is a "question mark."

A Critique of Terry Dosh's newsletter, "Bread Rising."

Most of the material in "Bread Rising" is edited material. Terry Dosh is Church historian. However, he focuses on many of the deviant, and wayward clergy. He is the head of CORPUS, the laicized, married priest movement. My criticism of "Bread Rising" is that it focuses too much on the drone relationships within Catholicism.

Patrick A. O'Dougherty, Ph.D.

120

IRISH PSYCHOLOGY/IRISH PSYCHIATRY: THE TIRESIAS COMPLEX

Bishop Larry Welsh: 6/27/95
Cardinal Dennis Dougherty and Cardinal William O'Connell
The New Deal("Raw Deal") Revisited
Cardinal Dennis Dougherty and Cardinal William O'Connell were famous critics of the New Deal("Raw Deal") National Catholic Welfare Council. They argued it was socialistic. This council later became the National Conference of Catholic Bishops and the United States Catholic Conference. I think that Dougherty and O'Connell were correct. They set a tradition which "shouldn't die."(Response: Charlotte O'Dougherty) Both the National Conference of Catholic Bishops and the United States Catholic Conference are largely "socialistic." (Ref. Margaret Mary Reher: Catholic Intellectual Life in America) These groups are "Writers on the Left."(Ref. a book by Daniel Aaron) The New Deal was a "Bloodless Socialist Coup." (Ref. Brother Gregory Conant, Saint Benedict Abbey, Still River, Harvard)
("A Bridge too Far")(WWII battle)
There are only around 70 million Catholics in this country. The government is asking the Catholics and Christians here to spend their lives bailing out failed government programs costing millions, billions, and trillions. I can't possibly advocate this. I don't mind carrying the cross, but I can't bench press a three thousand pound barbell. The government is an incubus or three thousand pound barbell on the chest. Catholics don't get much in their schools, on television, in the press, or in medical. Why be a chump? America has become "A Bridge too Far" for Catholics.
"Cut your losses short," (John O'Dougherty)

Patrick A. O'Dougherty, Ph.D.
Also, Have you read Susan Faludi's, Backlash. She is an Italian American thinker? America is a "Demented Inn." (Expression Thomas Merton used for the world) Instead of the term, slave, found in the thought of Saint Louis Marie De Montfort, I would use the term, freedman or freedwoman.

Bishop Lawrence Welsh: 6/30/95
St. Thomas College?
I came across a book by Lawrence C. Soley, Leasing the Ivory Tower: The Corporate Takeover of Academia. Is this true for the Catholic Universities? My uncle, John O'Dougherty, thinks that it is true for Father Theodore Hesburgh and his relationship with the Rockefellers. Is it dollars for dogmas at St. Thomas?
Terry Dosh's "Merry-go-round"
Richard Hope, John O'Dougherty, and Patrick O'Dougherty all have the same impression of Terry Dosh. Almost none of his material is original. He is just jumping on a "merry-go-round." Is that the role of a Catholic intellectual? Dick Hope thought I should share this with Terry Dosh, that is, see enclosures on Leonard Swidler's

121

IRISH PSYCHOLOGY/IRISH PSYCHIATRY: THE TIRESIAS COMPLEX

Catholic Constitutional Convention and the Catholic Bill of Rights). I call him Leonard Swindler--some good, some swindle.

Italy: Europe's House of Cards?
Problems in Italy: They Lost at Blackjack

First, I thought that the Italian people appeared more subdued when I visited them than the Germans, the French, the Austrians, the English, the Scots and the Irish. Why is this? Did somebody break a wine bottle over their heads? Second, I think Italy needs a face lift like Paris had under Malraux and De Gaulle. Third, is Italy the world's largest nudist colony? Fourth, is it a Marxist country? Fifth, the Italians elected Benito Mussolini's grand daughter on the Fascist ticket to the legislative body of Italy. To carry out her party and program, she is making pornography films with herself as the star. Is she trying to drop a blond bombshell on Europe? Sixth, under Prime Mininister Arnaldo Forlani (1981) many cabinet officials were found to be members of "an illegally secret Masonic lodge." (Ref.: World Almanac 1987) Is this conspiracy still in existence? Seventh, the situation in Germany is quite stable compared to the situation in Italy. Is Italy, as John O'Dougherty thinks. "Europe's House of Cards." Is it Europe's problem rather than its salvation? Eighth, is Italy a "post-Christian country?" Many analysts, for example, Thomas Vadney suggest this. Patrick O'Dougherty, Ph.D.

-----Father Hilary Freeman, Holy Rosary, thinks that the priests in the archdiocese preach "sociology rather than theology." He studied with Michael Dummett and A.J. Ayer at Oxford. He is a friend of mine. He deserves a good retirement position.

Bishop Lawrence Welsh: 7/3/95
Lincoln/Atheist or Agnostic

Abraham Lincoln was an atheist or agnostic. Some scholars argue the case that he was a Christian, but the argument is shaky. Lincoln wanted to be an Old Testament non-Christian leader.

Thomas Jefferson/Deist or Atheist

"To talk of immaterial existence is to talk of nothings. To say that the human soul, angels, God, are immaterial, is to say they are nothings, or that there is no God, no angels, no soul. I cannot reason otherwise."--Thomas Jefferson, in a letter to John Adams, August 15, 1820. (Anonymous source)

Catholics and Nationalistic Wars

One of the main acts that Orestes Brownson did after his conversion was to argue the case for Catholic involvement in the Civil War--a nationalist war.

There is only about one year of peace for every nineteen years of war in Western Civilization. (Source: John O'Dougherty) Why should Catholics fight in wars unless the Pope declares them just wars? The Vatican did not declare Vietnam a just war.

Israel: A Sellout?

IRISH PSYCHOLOGY/IRISH PSYCHIATRY: THE TIRESIAS COMPLEX

Israel has bargained land away for American dollars. If this happened to America, we would call it a sellout.

Hitler and Virginity

Hitler was a lifelong virgin. He wouldn't allow a fan to kiss him. Virginity was part of his race war fulcrum. He made the Jews and Marxists give in on race and class over the single virgin male soldier issue.(Source: David Wallechinsky, The Book of Lists, 1978)

Catholic Seminaries: Nietzsche and Heidegger

I read in Fidelity Magazine that Nietzsche and Heidegger are all the rage in many Catholic seminaries. Nietzsche was an "autocratic elitist." Heidegger was a Fascist.

St. Catherine's College

I heard that the women are thinking about changing the status of St. Catherine's to nonsectarian. You never hear much about St. Catherine's in the press. Why is this true?

Patrick A. O'Dougherty, Ph.D.

Bishop Lawrence Welsh: 7/5/95

Cathometrics: A New Word I've Coined

The exploding areas of research in history are psychohistory, cliometrics, urban history, and the history of science. These areas are all going to become major areas of Catholic analysis and Catholic history. The psychology of the Catholic, Cathometrics, the urban Catholic and Catholics in the history of science are the crest of the present and future. Intellectual history and radical history are somewhat older fields. The idea is to pin down concepts with analysis in these fields. For example, Leonardo Boff, the Brazilian, does not try to pin down his concepts. However, Catholicism has traditionally favored a more rationalistic approach. Have you ever been analyzed?

Patrick A. O'Dougherty, Ph.D.

Alphonse J. Matt Jr.
Editor, The Wanderer
                            8/23/95
Dear Mr. Matt,

Problems at Newman Center Delineated

IRISH PSYCHOLOGY/IRISH PSYCHIATRY: THE TIRESIAS COMPLEX

I am responding to a letter written by you to Most Rev. John R. Roach dated August 18, 1995. I am enclosing a copy of the Friday, September 23, 1994 Minnesota Daily article on "Problems at Newman Center." It outlines several of the major problems at Newman Center. As Mrs. Cindy Paslawski suggests in a letter dated August 10, 1995, "it is a situation in which their own actions will condemn them." Another major problem that has arisen at Newman Center is the problem of Catholic Marxism. See the letters on Catholic Marxism and other issues which I sent your staff last Friday. At Newman, they have Brechtian liturgies. He was an East German Marxist dramatist. The Pathfinder Bookstore, a political socialist bookstore/movement, was given a propaganda table at Newman Center. A Sandinista leader was given a seminar in the basement of Newman Center. I made a case at Newman Center with Father Steve Bossi that the Church is about everybody. Newman Center should open itself to many perspectives within the Church. He finds this to be unacceptable and thinks it would divide the congregation. He meant financially. Economics are the big problems in life and around which most issues in adult life revolve. (Sources Aquinas O'Dougherty and Shirley Corrigan). The main problem at Newman is money.

A Major Catholic Philosophy and Major Research Projects at Newman

I have been a longtime member on an off at Newman Center since the 1960s and an activist for some years with the International Alliance for Sustainable Agriculture located in the Dorothy Day Room. I have built with several others an internationally recognized agriculture research library there. I am sending you a list of some of the material in this library and a copy of a letter to Dr. Terry Gips of my research interests at IASA especially in Hydrogen Conversion. **I have invented a major intellectual position in Catholicism: personal, intuitionism. It relates Dorothy Day's, Scheler,s and the Pope's personalism to mathematical intuitionism.** I feel this is much more acceptable than dated Marxism. I have received several positive cards from the bishops in the chancery. Archbishop Roach, for example, told me in a card that he found some of my ideas to be interesting.

Money Problems at Newman and in the Archdiocese

The main problem at Newman Center is that the center is failing financially. Father Steve Bossi attempted to turn the center around, and he drew a line with me over the money problems and expanding the intellectual forum of Newman Center. I have never done that. Is this acceptable behavior for a priest? I sent the Editorial Staff at The Wanderer several letters that I have written the bishops on the budget request Newman made of the archdiocese for $199,000. Does The Wanderer get that from the archdiocese? I feel that the center has become a largely political, rather than spiritual center. Also, there was a check theft inside job robbery at Newman Center. They did not handle it properly. I wanted to make sure they filed charges so the individual would not comeback or retaliate at the International Alliance for Sustainable Agriculture. The authorities did not apprehend the individual involved, but his bullet ridden car was found in Hopkins. Moreover, Newman Center is not a popular center, like St. Joan of Arc or St. Agnes Church. There have been a lot of major financial losses at Newman Center and also in the Archdiocese of St. Paul. I think the main reason is freedom of ideas. There are two major ways to make money "hard work or hard thinking."(Source Aquinas O'Dougherty) I think that the freedom to agree or dissent within the Catholic identity expressed within the Land O' Lakes Magna Charta Statement is causing the financial problems in the archdiocese. Ideas are the main solution to the economic problems in the archdiocese.

The Comfortableness Issue

Alphonse Matt sent me a letter from Archbishop John Roach dated August 15, 1995. In it he raises the issues of **pastoral procedure** and the **encouragement issue**. I would like to head these issues off a little bit by raising the issue of comfortableness in relationships. I think that the problem is the archbishop does not feel comfortable with me. This is a big problem in any administration. I am not a priest, or theologian, or on his payroll. I am an intellectual historian. Many people do not particularly like intellectuals. However, many people do not like priests, or bowling professionals. He was the head of a military academy. Maybe he doesn't like dissidents. At Newman Center, I think they want me to test the waters with the archbishop. I have always liked Archbishop Roach, but I don't feel comfortable with him. Other people in the archdiocese have expressed this same concern to me. For example, Jacqui Landry told me that the reason I have not been given any recognition by the **current** staff at Newman "is because they don't get any." Ted Hartman, deceased, of Holy Name Parish told me that he thought Archbishop Roach had too much "cronyism" in his major appointments. However, in fairness Ted didn't get one. Louise Hall, Legion of Mary, asked me to take on the staff at Newman Center for not addressing the atheism/secularism on campus. She thinks the archbishop is "vacillating." Father Hilary Freeman raised the issue of modernism in the next papacy with me. He told me that St. Catherine's College has become "the history of scandals." Dr. Mike Franey, a former student of Archbishop Roach, thinks that the archbishop does not go for Alphonse Matt and The Wanderer. I have received positive letters from the archbishop and Mike thinks the response from Cindy Paslowski "piqued" him. The Cathedral does not carry The Wanderer. In relationships, if you don't feel comfortable you shouldn't do them. I have been a counselor at the university. This is an area I know something about. However, the Church teaches humanity has a fallen nature. For example, there are a lot of frustration, rejection and problems involved in dealing with people. This is why I focus on ideas. The key to relationships is flexibility. I hope the archbishop eases happily into a well deserved retirement. He deserves a "last hurrah."

IRISH PSYCHOLOGY/IRISH PSYCHIATRY: THE TIRESIAS COMPLEX

Patrick A. O'Dougherty, Ph.D., personalist, intuitionist.
cc Archbishop John Roach, Bishop Lawrence Welsh, Bishop Harry Flynn, Mrs. Cindy Paslowski

Archbishop John Roach: 9/8/95

Request for a Change of Venue
An American Raises the Europeanist Controversy

I have presented my case to you, and I am formally asking for a change of venue to other archbishops. You were the head of a military academy, and I don't know if I would get a fair hearing from you. Nearly 120-130 million people perished in Europe in this century. Several different thinkers have come up with this figure. It is still conjecture. However, millions of people in Europe starved to death for military expediency, then the military leaders declared there was a famine. John O'Dougherty told me that the figures for one extended battle on the Western front in WWI were one million dead. In contrast, only twenty million buffalo died during the period 1850-1889. (Source: World Book Encyclopedia, 1995) Europe is a culture of war lovers, war is being used for population control (Sources: Sheila Franey and Patrick O'Dougherty), fallen human nature is given free reign, and this the way they raise children in Europe. There is something "rotten in Denmark" when death figures this high come out. This is the biggest lie or deception that told the masses in Western Civilization. Most of them died for nothing. This is an absurdity. Edward Haugh told me that the Blessed Mother revealed to Jacinta at Fatima that most of the people who fought in WWII went to hell. This is true for both sides. In contrast, Christ was never a soldier. However, he never opposed the military. He accepted capital punishment. Moreover, I cannot possible imagine that he would not formally condemn political and military excesses and political and military leaders like our present day ones. John Harris thinks military "diplomacy is based on deception." I am not deceived. Several people have raised this issue with me.

Sanctuary for the masses,

Patrick A. O'Dougherty, Ph.D.

Archbishop John Roach: 9/21/95

"A Farewell to Arms"

126

IRISH PSYCHOLOGY/IRISH PSYCHIATRY: THE TIRESIAS COMPLEX

I would like to congratulate you on your retirement. You deserve it. I thought you were an excellent priest, bishop and archbishop in our archdiocese. I hope many more priests come from the seamless bolt of clerical cloth that produced you.

On Newman Center

Do you think the staff at Newman Center is treating me in a Catholic or Christian manner? Many people, including Mark McGee, do not think they are? I want the issues resolved there. I did not start the fight there.

A Right to Protest

Here is St. Patrick's, LETTER TO THE SOLDIERS OF COROTICUS. It details PATRICK'S RIGHT TO PROTEST.

A Prayer for a Bishop

Enclosed is a "Prayer for the Bishop."

Pray for the wild clover,

Patrick Aquinas O'Dougherty, Ph.D., the wild clover

Archbishop Harry Flynn
Bishop Larry Welsh
Archbishop John Roach

12/14/95

The Convergence of the American Experiment and Marxist Experiments

I want to affirm the tradition of Orestes Brownson, John Ireland, Cardinal Dennis Dougherty and William Buckley in American Catholic history. I am going to throw the massive hydrogen conversion project and the alternatives to petro-chemical to this tradition. There is a sixty-five year convergence between the American experiment and the Marxist experiment. For example, the two parties in America, "Tweedledum and Tweedledee," form a phony dialectic. I wrote my seniors honors paper on this topic back in 1968.

IRISH PSYCHOLOGY/IRISH PSYCHIATRY: THE TIRESIAS COMPLEX

Also, John Marty, Martin Marty's son, is leading the fight against school vouchers in the Minnesota Legislature.
Patrick O'Dougherty, Ph.D.

Archbishop Harry Flynn
Bishop Lawrence Welsh
Archbishop John Roach

Bishop Lawrence Welsh: 2/13/96

Rosie's: A St. Paul Cathedral Neighborhood Lesbian Bar
No Man's Land
I used to live in the Cathedral neighborhood. I had a Catholic friend named, Mike Smith, who liked to shoot pool. There was a bar on West Seventh, named Rosie's, just a short distance from the present day Day's Inn.   The bar was previously named Jack's Bar. We went into to Rosie's one summer evening to shoot pool. It was wall to wall lesbians. Rosie's is just down the back door from the Chancery and the Cathedral. The lesbians bought Jack's Bar to introduce a counter culture to Archbishop John Roach our friendly priest from Prior Lake. Are the themes of Jack's and Rosie's, Kennedy themes?

Edgar Allan Poe and Archbishop John Roach

Another act that happened during Archbishop Roach's tenure was a woman was murdered right by the exit to the street car tunnel that comes out from under the Cathedral. This is practically on the doorsteps of the Chancery. Edgar Allen Poe, a West Point Academy dropout, thought the most powerful themes in literature are death and beauty. He thought the death of beautiful woman unites these themes and creates the most powerful statement in art. (Reference: Clifton Fadiman, The Lifetime Reading Plan) These themes come together tragically in the murder of the beautiful woman next to the Chancery. Thus, Poe transfixes Archbishop John Roach's tenure in St. Paul. History will reflect on Archbishop John Roach and the lost world of Archbishop John Ireland.

Patrick A. O'Dougherty, Ph.D.

Archbishop Harry Flynn: 3/24/96
A Peace Curriculum in the Catholic Schools?
How would you initiate a peace curriculum in the Catholic school system? This curriculum should look at peace in theory and in praxis, and it should have a scientific and mathematical basis.

IRISH PSYCHOLOGY/IRISH PSYCHIATRY: THE TIRESIAS COMPLEX

Siena College: The Sustainable Agriculture Program
I made an initiative with Brother Gregory Conant, OSB, about starting a department at Catholic Universities to address issues in sustainable agriculture. He contacted the president of Siena College about this proposal and a department or program was begun in the field of sustainable agriculture.

Patrick A. O'Dougherty, Ph.D.

Archbishop Harry Flynn
Bishop Lawrence Welsh
Archbishop John Roach

Excellencies: 2/22/96
"The Gay Deceiver"
Jeffrey R. Anderson, the attorney, E. 10000 National Bank Building St. Paul, who is getting all of these massive legal judgements against the Archdiocese of St. Paul for victims of homosexual priests in this archdiocese, is himself a homosexual. Did you ever hear of homosexual pyramiding in the military? Homosexuals are not necessarily mentally ill. However, they are often psychopathic on the MMPI, meaning very manipulative. For example, Alfred Kinsey who spent much of his adult life studying the sex habits of young boys was also a homosexual. This came out after his death.

Patrick A. O'Dougherty, Ph.D.

Father Hilary Freeman:
Brother Gregory Conant:
Sister Mary Anthony Wagner:

                                  2/25/96
Patrick A O'Dougherty: A black sheep, A "white Negro"
2/5/7 A black sheep goes to heaven
I was born in 1946 shortly after WWII. I grew up in the 1950's in South Minneapolis. I got published in grade school. My dream as a youth was to address intellectually the contribution made by the Jews in Europe in the holocaust and their contribution to materialism. Also, I had a nun in my grade school, St. Thomas the Apostle, who gave us the statistic that the Belgian Congo was 99% illiterate. I wanted to produce a very formidable intellectual "black classic" to brace the experience of the black people world wide and to help pull them out of illiteracy. This is still my goal. The tools I have chosen to define this mission are wanderer, Catholic dissident, and psychological analysis. A concept used to integrate the

experiences of my life is "Mestizaje." If you should die, I wonder who might be willing to work with me intellectually at St. Thomas Aquinas Priory?

Back the Green, the Grey and the Purple Viconian Revolutions,

Yours in Christ,

Patrick A. O'Dougherty, Ph.D., time wanderer

Father Hilary, I hope you do write a short book on philosophy. I think the field of American Catholic philosophy is in bad shape. The key to Catholic philosophy is Freeman. The answer is also Freeman.

Brother Gregory, what are your ideas on hydrogen "conversion?" I found out that desalination is not a big problem in tapping the hydrogen in seawater if one uses electrolysis. If you should die, who is someone at St. Benedict's Abbey who want to do something formidable intellectually.

Archbishop Harry Flynn: 3/14/96

The Maranatha Bookstore

I like this bookstore. I encourage their efforts. However, I give it a mixed review. Most Marian devotions are private devotions. John O'Dougherty thinks, "they mean well." However, one cannot accept visions of the Blessed Mother as commonplace.

Patrick A. O'Dougherty, Ph.D.

Father Mike Tegeder: 3/23/96

Private and Confidential

This is something I have filed with different lawyers.

I am a writer, an activist, and a psyche consumer. I got a psychologist, named Dr. Shirley Corrigan, while I was an undergraduate at the University of Minnesota. I continued this course until the present day. The deal is if you are a writer you can't just learn from textbooks or experts.   You have to amplify experiences. Also, you need to have an indeterminate position in society for literary freedom. Instead of seeking a corporate job, I walked out of the door in graduate school with very little money. I hitchhiked down to Atlanta, Georgia, to see what would happen. A lot happened. Anyway I ended up getting processed by the courts and the social agencies. I have always had a normal profile on the MMPI. I ended up getting tardive dyskinesia from Prolixin. I had this problem for eleven or more years. Dr. Patrick Stokes and Dr. Joseph Gendron were responsible for this drug prescription. I suffered horribly from this drug with psychic dependence, tremors, pseudoparkinsonism, falling, inability to get deep sleep, and almost daily vomiting.

IRISH PSYCHOLOGY/IRISH PSYCHIATRY: THE TIRESIAS COMPLEX

Under Dr. Joseph Gendron, I ended up having major surgery on my knee from falling three years ago. He removed the drug and put me on Thorazine. This drug enabled me to sleep. However, I feel there is a social control issue here with Prolixin. I raised the issue of major surgery with Dr. Gendron and the tardive dyskinesia contraindications. He decided to discontinue his outpatient practice and semiretired rather than deal the legal problems posed by this suffering. He gave me some psychiatric referrals. What are my legal and patient rights in a situation like this? What are my rights in the area of literary freedom as a writer and psyche consumer? Could you delineate them for me? Kate Millett was another local writer and psyche consumer. She addressed her problems in a book, The Loony-Bin Trip. I've had enough suffering. How do you evaluate my intellectual, religious, social, and political contributions coming out of a psyche consumer experience?
A pedagogy for the psyche resurrected,

Patrick A. O'Dougherty, Ph.D.

Bishop Lawrence Welsh: 3/25/96
A Catholic Scientific Black Classic
I worked with Brother De Paul in his mission to Haiti to build schools and develop sustainable agriculture in their country. I sent money to a seminarian, named Juan Diego, in a poor part of Mexico. I worked with Father Terry Dougherty and Father Robert Monaghan at Incarnation. Father Terry Dougherty is a missionary to Monrovia and Liberia in Africa. Also, I have watched people starving and in need of medical attention in Latin America. **As an intellectual, I am trying to address the contribution made by the Jews in Europe to materialism and to address the starving and illiterate masses of the world with a massive Catholic scientific synthesis. I got a doctorate in intellectual history and the history of science to further this childhood goal and dream of mine.**
End of the case, Patrick A. O'Dougherty, Ph.D.

Father Hilary Freeman: 3/26/96
Mathematics and Deep Ecology
What powerful mathematical constructs would you relate to the ideas of mystery in nature, hierarchy in nature, connectedness in nature, the human, complexity, integration and entropy in nature? I am going to develop the theme of changing chaining in nature to bonding. This is the theme of St. Patrick and Patrick O'Dougherty. What mathematical theme describes the theme of freedom and freeman well?

Patrick A. O'Dougherty, Ph.D.

IRISH PSYCHOLOGY/IRISH PSYCHIATRY: THE TIRESIAS COMPLEX

Sister Mary Anthony Wagner: 3/29/96
"Pat's City"
The McKnight Building Riverside Plaza: Mary Tyler Moore's Building 1615 South Fourth Street

I live in the building in Riverside Plaza that Mary Tyler Moore moved into when she moved from her home near Lake of the Isles. She is a different type of woman than Sister Mary Anthony Wagner. I have a penthouse apartment on the thirty-fourth floor facing south. I am sending you a brochure on the building. There is a free school here though tenth grade. It has a free community room on the plaza level where we could have the Oblates meet. I really don't cook anything, and I eat all of my meals out. We could have a breakfast or lunch in one the many ethic restaurants in the neighborhood, then we could have our meeting in the community room. We could also have rolls, however, the restaurants are more fun. For example, there is an East African Restaurant across the street named the New Chile Time. Pat Miles gave it rave reviews. Also, Jerry Petermeier who went to St. John's University and got an MA in Theology and MA in philosophy from St. John's owns The Wienery right on Cedar. He cooks gourmet wieners. He is from Melrose. There is also a vegetarian restaurant called the Riverside Cafe which has very good ambiance for young students. There are many other restaurants, especially oriental restaurants, in this neighborhood. None of these restaurants are very expensive. The parking is also inexpensive. Riverside Plaza, a suburb within a city, overlooks I94 which goes right by St. Benedict's College. I had the Legion of Mary women meet here, and they liked it very much.
Happy Easter,

Patrick A. O'Dougherty, Ph.D., Oblate, wanderer, publisher, freeman

Bishop Lawrence Welsh: 4/4/96

The Rights of Single People: The American Majority
Common Sense III
I am not going to be for a married clergy movement like Terry Dosh is advocating. Maybe I was too tough on Terry Dosh in the letter I wrote you and Terry. However, forty percent of young people today are living with fathers who are not their biological fathers. If the door opens to a married Catholic clergy, we would end up having children who survived abortions, surrogate clergy mothers, and every other type of arrangement in the clergy. A single person, like myself, would be paying for these situations.
Tough Stuff

IRISH PSYCHOLOGY/IRISH PSYCHIATRY: THE TIRESIAS COMPLEX

I have worked practically everyday since I was eleven years old. I committed academic suicide in ROTC at St. John's over the war in Vietnam. I became a charter member of S.D.S. and involved myself in the Young Socialist Alliance. My father, Aquinas, wanted to get me toughened up. So, he let me take an eight month sentence on Bellwood Chain Gang in Atlanta for practically no offense. However, I am a psyche consumer. I went to Northern Ireland during the Troubles. I am single and celibate. I am a founder of the Catholic Green Revolution.

The Catholic Green Revolution: Through a Glass Darkly

I am part of the Catholic approach to the Constitution for the Federation of Earth and the World Parliament. I am for the Green Party, Green Caucus internationally. Petra Kelly founded the Green Party in Germany. The Germans are tiring of the Red Revolution. Back the Green Revolution, an ethical, scientific, theological and political revolution! Look at Catholicism and this movement through a glass darkly which is a theme from St. Paul. Father Hilary Freeman suggested this theme to me as an approach to liturgy.

Pro life all species, Happy Easter,

Patrick A. O'Dougherty, Ph.D.

Father Hilary Freeman: 3/31/96
Praxis

I am sending you a copy of a movement for the federation of the earth and the world parliament movement. Consider it an Easter gift. What is your reaction to these ideas? I am going to use it as the praxis for my ecology book. I am going to include my library as a data base for the book. What mathematical ideas or logic ideas might apply to ecology? I will use the articles that you sent me.

Patrick A. O'Dougherty, Ph.D.

Sister Mary Anthony,   5/2/96
"A Black Classic"

I guess you don't feel real comfortable with me on the personal revelation or research level. The deal is I don't really know that much about women. I'm single. I've never married. I had a young manhood when I was in college and graduate school. Is that too much for you to handle? I never really went out with a girl from St. Benedict's when I was at St. John's University. What would happen if I went home to meet a St. Benedict's girl's parents in a town like Hibbing or Monticello or Miesville and told them that Pat wanted to do a "black classic" and go through a whole lot of changes in life, for example, revolutionist, psychoanalysis, and Catholic Republicanism and stick with the Catholic faith? My father was from Maple Lake and

IRISH PSYCHOLOGY/IRISH PSYCHIATRY: THE TIRESIAS COMPLEX

he could have opened a law practice in a small town in Minnesota. However, he didn't do this because he had a son that wanted to go through a variety of intellectual changes. Did you ever go to the scene of a terrorist bombing on a date? I did this once. It is not a sin to do this. I like the nuns.

The 1000th Year of Austrian History is This Year
This is the 1000th year of Austrian history. I have been with the Austrian Studies Program for eleven years, so I thought I would write a work in liturgy. Do you have a reading list in liturgy that you could send me?

"Dogs, Irish and Negroes Do Not Enter"
Back in the 1950s they had signs throughout New England saying the above statement. They still have signs like this in some English pubs. Like women and children, the Irish really don't have many weapons. You lived through the "dark ages," the 1950s. These were the "dark ages" for minorities, women and children. Aren't you glad that era has passed? Don't you think these Americans need to get toughened up spiritually, intellectually, and psychologically?
St. Benedict's, Through a Glass Darkly,

Patrick A. O'Dougherty, Ph.D.

Bishop Lawrence Welsh: 6/15/96
Getting a Handle on Newman Center
I am a writer, activist, businessperson, client of psychoanalysis, professor, devote Catholic, Newman Center member since the l960s. There was a change of staff at Newman Center five years ago with Father Steve Bossi taking the helm. He is a military man out of the South, and I'm a dissident. The center wasn't doing well financially. So Father Steve Bossi drew a line with me over the money issue. I get to have no rights, no say, no respect at Newman Center. I appealed to Archbishop Roach about what to do about this and he took a well deserved retirement. I grew up in this area and I do not want a long term negative relationship in my life. I didn't know what to do about this so I called a black attorney at the Human Rights Department about this. I took his advice and called the City Attorney about it. He told me what to do. I did it. I also filed this situation with a Catholic woman psychologist and took her advice. She thinks the problem is Father Steve Bossi is too Bossy. I also talked to Father Mike Tegeder about this.
St. Mary's Hospital
I also ended up having unnecessary major surgery at St. Mary's Hospital. I thought I would throw this in Father Steve's ballpark to see what he would do with it. He has a major neurology problem, and he ended up having a fall, and shortly thereafter a car accident with Newman Center students on a trip. My Father, Aquinas, ended up getting a staphylococci infection at St. Mary's Hospital after a

minor surgery. He almost died. I would have grown up without a father and been unable to attend St. John's University. My uncle, Terrence, ended having major surgery repeated five times at St. Mary's Hospital. He died from pancreatitis resulting from the last surgery. Nothing was done about this because the doctors subtly told us to "turn the other cheek." Would you go for this?

Americans with Disabilities Act

Robert Dole sponsored this act. Bill Clinton signed it. I want to see a major Catholic disabilities archive and focus come out of this act and in this archdiocese. Maybe somebody might do a longitudinal study and find a say genetic cure for priestly alcoholism. Let's put this in Bishop Larry Welsh's ballpark. I have always wished the clergy well.

You can get something on anybody, right,

Patrick A. O'Dougherty, Ph.D.

Father Virgil O'Neil,: **"Through a Glass Darkly"**

When I dropped out of St. John's University in the 1960s, I reinvented the prodigal son. I ended up becoming an Irish Catholic Viktor Frankl kind of a man. You are a great man and a great priest. I worry not about your brother out in Idaho. I worry about you as a Benedictine. I am sending you an update on my life and research. I think that you should consider looking at the whole Benedictine tradition as St. Paul would "through a glass darkly."
"Listen, my son," God Bless,

Patrick O'Dougherty, Ph.D., black sheep, true believer, Benedictine Oblate, Irish Catholic white Negro.

Father Virgil O'Neil: 6/25/96

Catholic Personalist Intuitionism

I have come up with a new school of thought in philosophy: the Catholic personalist intuitionist school of thought. I have plugged Jan Brouwer's intuitionist or activity school of mathematics into Dorothy Day's, the Pope's and Scheler's personalism. Moreover, numbers are internal personal activities. The person integrates mathematics. Mathematics is personalism. I am sending you some material on this school of thought which I created out of the holocaust experience, the Civil Rights Movement and my Irish white Negro and Viktor Frankl experiences. What do you think of this school of thought? Politically, I am a theocrat, an aristocrat and a radical republican like Thaddeus Stevens of Pennsylvania and Charles Sumner of Massachusetts who stood up for the black people during Reconstruction.

Patrick Aquinas O'Dougherty, Ph.D., St. Thomas Aquinas was an intuitionist.

Father Virgil O'Neil: 7/6/96
A Devout Catholic
I have been a practicing Catholic most of my life. I go to daily Mass during the school year. I receive the other sacraments regularly.
War Culture
There have been 125 million war related deaths in Europe so far this century. What do you think of a figure like that? Do you think the Jews and the Slavs are subhuman peoples? This was the propaganda during WWII. I don't feel comfortable with this.

The Windsor Family: Dumb English/Germans
My mother and my uncle John do not feel comfortable with the Windsor family. Princess Diana's aptitude scores were so low that they weren't going to let her graduate from high school. So, she has an I.Q. below 100. The English are the big losers in Europe. How many civilizations have lost that badly? The Tudors and the Plantagenets were the strong families in English history. The Windsors are very average. My mother told me that all Elizabeth talks about is horses. She got me a scholarship for the Austrian Studies Program. Austria is a better bet than England. My book on liturgy is going to commemorate the 1000th year of Austrian history.

Patrick A. O'Dougherty, Ph.D., personalist/intuitionist

Father Virgil O'Neill,   7/13/96
Ordination Jubilee
I congratulate you on your Ordination Jubilee. I like the Germans all right but an English/German, family, the Windsors, are killing Irish Catholics in Northern Ireland. The Windsor family are lowbrow. I decided to move on from teaching in General College and become a Catholic revolutionist with a scientific bent. People who join radical religious or extremist political groups like your brother Dick score high on the F-scale on the MMPI. Have you ever thought about publishing something? I wish I had your ability in music. My brother, Mike, has a pump business in Benson, MN. You have had an outstanding career. What are you going to do to cap it off?

IRISH PSYCHOLOGY/IRISH PSYCHIATRY: THE TIRESIAS COMPLEX

Also, I'm trying psychoanalysis and analysis.
God Bless,

Patrick O'Dougherty, Ph.D.

Sister Mary Anthony Wagner,   8/18/96
A Major Work in the Women's Movement
I wrote a major work in the women's movement which I would like you to
read. It is Personalism and Mathematics as Women's Personifestoes. The idea of
it is to counter the Communist Manifesto with the philosophy of personalism and the
intuitionist school of mathematics created by Jan Brouwer. In the work I use Bible
passages to prove the intuitionist school of mathematics. I developed a new school
of physics out of the holocaust and the works of Lise Meitner and Edith Stein in the
book. Several of Michael Dummet's Oxford lecture notes are in it. He followed
Bertrand Russell at Oxford and worked with minorities like I have. I want to apply
this philosophy to psychology and anthropology and to the Viktor Frankl experiences
I have had in life. I have moved on from General College and have decided to
become a full time writer, activist and revolutionist.   They should have the
Benedictines at the University of Minnesota. I think that I've opened the door at the
University of Minnesota to the Benedictines. Also, how does St. Benedict address
personalism, mathematics, "meditation in motion" like Tai Ji and group therapy.
Up from Bondage,

Patrick A. O'Dougherty, Ph.D., Irish white Negro

Father Hilary Freeman, O.P.   9/5/96

My book on St. Patrick, The Green Revolution, and the Hydrogen Conversion
Project is almost done. Thanks for your suggestions. I have been thinking about
writing a book on the letters of Patrick O'Dougherty. I have nearly 400-800 letters
in my files in my computer. There are many topics in these letters ranging from
Duncan Kennedy's Critical Legal Studies to Issues at Newman Center to
Psychoanalysis. I am considering using a conceptual framework of Christian freedom
to organize or preface these letters. Do you have any suggestions or references on
this topic from your perspective as a philosopher?   I am going to term the era in
which I wrote these letters my Blue Period to parallel Picasso's Blue Period in art.

God Bless,

IRISH PSYCHOLOGY/IRISH PSYCHIATRY: THE TIRESIAS COMPLEX

Patrick A. O'Dougherty, Ph.D.

Father Hilary Freeman: 2/15/96

I have developed my own school of Catholic thought, personalist intuitionism. I want to apply it to liturgy. I am sending you some material on Wittgenstein. How would you apply his thought to liturgy? I am sending you some material developed by Tom Conry on "Looking at Liturgy: Aristotelian VS. Brechtian Lenses." How do you rate these lenses to look at liturgy from the perspective of Wittgenstein? Do you agree with these lenses? I will apply my school of thought to analyzing all of these approaches.

The reason I am writing a book on hydrogen conversion and another book on alternatives to petrochemicals is to address intellectually the contribution made by the Jews in Europe to materialism. These are also cutting edge issues in economics and medicine. Also, the American capitalist system of economics bases itself on "self-interest," materialism and greed. Do you find these themes to be acceptable for the Christian? Perhaps this might be another book. The basis of economics is the Freeman. Catholic philosophy is Freeman.

Patrick A. O'Dougherty, Ph.D.

Father Hilary Freeman:
Brother Gregory Conant: 3/8/96
The Definition Problem and Darkness
The way to define the word priest is: Father Hilary Freeman.
Thanks for your essay on Wittgenstein. I will use it. This is the 1000th year of Austrian history. I address it from the perspective of liturgy. However, I am not a priest, theologian or liturgist. I don't want to cross wires with the clergy or do something blasphemous here. Should I write a perspective, interpretation, or approach to liturgy? What would be an approach to liturgy that a clergyman might feel comfortable with? I am sending you some material from St. John's University on liturgy "as a Welcome Place for the Displaced," the idea of stranger, Benedict's Cave, Mestizaje, Freedom, Creation, Exodus and centering prayer. How would you define these terms darkly?
"Through a Glass Darkly": A Powerful Theme
Do these words describe God? Are they analogies for God? In logic, analogy is one of the weakest forms of proof. Is darkness a form of proof? As you mention, Wittgenstein's notion that things before our eyes escape our notice and description must take the place of explanation. How would "looking through a glass darkly" affect the description of the words that we apply to God strictly, for example, goodness in God or liturgy? How would you define liturgy looking at it "through a

glass darkly?" This is a quite powerful theme you've chosen from St. Paul. How would you relate it to Mestizaje? What is an effect? I am sending you some material from a Benedictine College product, Allen Janik. He is trying to come to terms with the materialist conception of history. For example, the problem of the "relationship of ideas to ideas and antecedents to consequences." How would you look at the term, effect, in this context, for example Heisenberg's uncertainty principle, black body radiation, mysteries, and the social context of "culture" like the displaced in liturgy? How would you look at the term, effect, through a glass darkly?

Women: "Through a Glass Darkly"

You taught at a women's college so I'm sending you some material on "Austrian women in the nineteenth and twentieth centuries." How would you see women "through a glass darkly?" I hope you find it interesting.

Salman Rushdie: The Satanic Verses

After the recent suicide bombings in Israel, my uncle, John O'Dougherty, suggested I take a look at Salman Rushdie's, Satanic Verses. The author published it in London. In it Mahound is "the recipient of the revelation in which satanic verses mingle with the divine." Also, the Angel Gibreel at the end of the book put "the barrel of the gun into his own mouth; and pulled the trigger; and was free." The grey reality is often the Semitic one. Is this looking at the world "through a glass darkly." How would you rate a piece of literature like this one? Is suicide an act of freedom? I would appreciate a clergyman's analysis of a work like this. Moslem clergymen put a contract out on Salman Rushdie's life for writing this work. Did Catholic clergymen ever do this?

The Jews: A Curious and Curiousier Set of Facts

I had a paper route when I was a teenager in Minneapolis near 38th and France Avenue South. The irony is that the Jews in that neighborhood would give me larger Christmas tips than most of the Christians. They proved to me "through a glass darkly" that they were the chosen. They often loved one another more than the Christians in that neighborhood and they made much better parents. How would you rate experiences like these from your perspective as a clergyman?
Through a glass darkly,

Patrick, a single person, like St. Paul

Brother Gregory Conant, How would you respond to the above essays from your perspective as a Benedictine monk though a glass darkly? Ken Conant got the book, Life Culture Versus Death Culture and the Death of Literature. My sister, Mary Ann O'Dougherty, released a compact disc Water Color of Brazil. It is Brazilian jazz. Her band's name is Mandala. The music producing company is Deep Blue Records which is near the Conant neighborhood in Mpls. The price is $11.99 plus tax. Are you interested? Hollywood made a movie, Lilies of the Field, after a segment of my father's life. He was the handyman for the German nuns in the film. One summer while he was an adolescent he built a barn and chapel for some German nuns in Colorado. A writer wrote up the script about the story and it became a famous movie, Lilies of the Field, starring Sidney Poitier. Did you ever see this movie? Happy St. Patrick's Day!

Father Hilary Freeman:
Brother Gregory Conant: 3/16/96
Cathecosophy: Catholic Environmental Philosophy
        The way I am going to address the "hydrogen conversion" project is to look at the philosophy and practice of the new environmentalism, for example, the Deep Ecology Movement. I am sending you some material on it from George Sessions book, Deep Ecology for the 21st Century. He is chairperson of the Sierra College philosophy department in Rocklin, California, and coeditor of Environmental Philosophy. This movement has coined the word, Ecosophy, combining ecology and philosophy. I invented the word, Cathecosophy, to embrace this movement. It pulls the whole context of Catholic theology into the ecology movement. One basis of Catholic theology and ecology is St. Francis who argued for "'the equality of all creatures.'" Another basis is St. Patrick who rose from slavery to spiritual freedom or from mathematical chaining to bonding in the universe. A third thinker is Chardin who argued for the manifestation of the "'Cosmic Christ'" in evolution . A fourth thinker is St. Thomas Aquinas who argued "sacred doctrine is a science" which has a mathematical basis. The whole scientific paradigm and methodology relates to the ecology debate via the scriptures. A fifth thinker is Augustine who invented systematic theology is Western Civilization. Is nature systematic? Is theology systematic. I don't know if you can get a hold of George Sessions book, Deep Ecology for the 21st Century. You will find it interesting. How would you pull the whole ecology debate into the context of Catholic theology and the scriptures and into the hydrogen conversion project? Ninety percent of the universe is hydrogen. Is the word, sun, related to the word, son, for example, the Son of Man? The spring of nature, Lourdes, is the Blessed Mother. Also, animals do not live for the common good. What are some powerful methodologies to open the ecology movement?

Patrick A. O'Dougherty, Ph.D., the Godfather of the Ecology Movement

Sister Mary Anthony Wagner,   9/6/96
Confidential
        Maybe you should keep the confidence about counseling confidential. I also see a Catholic women psychologist. I find counseling and therapy to be a good frame of reference for a single writer like myself.

IRISH PSYCHOLOGY/IRISH PSYCHIATRY: THE TIRESIAS COMPLEX

The Blue Period Book

I am thinking about organizing my letters into a book. They cover several topics like Critical Legal Studies, Issues at Newman Center and psychoanalysis. I thought I might use Christian freedom as a preface or conceptual framework for the book. Do you have a reference to an original approach to Christian freedom? I also want to start a Catholic/scientist/activist movement. Amnesty International which you are in is an excellent form of activism.

Patrick A. O'Dougherty, Ph.D.

Archbishop John Roach
St. Thomas Academy
949 Mendota Heights Rd
Mendota Heights, Mn.

9/14/96

John,

John Kennedy and Robert Kennedy were Catholics silenced in America. Now the Republican Party has silenced Patrick Buchanan a loyal Roman Catholic. Am I supposed to go for this? I am voting for the Green Party and the Reform Party in this election.

God Bless,

Patrick A. O'Dougherty, Ph.D.

Sister Mary Anthony
St. Benedict's Monastery
104 Chapel Lane
Joseph, Minnesota 56374

Sister Mary Anthony,   11/7/96

I am writing to tell you that I like getting your Oblate letters. They are informative about issues and people in the monastery as well as about anniversaries, birth, and deaths in Benedictine and Oblate communities. I like the way you fit in the guide to spiritual readings. Your letters contain much detail which is a sign of intelligence; and, they also contain poignant spiritual anecdotes, for example the

141

IRISH PSYCHOLOGY/IRISH PSYCHIATRY: THE TIRESIAS COMPLEX

quote from Viktor Frankl, a sign of quality writing. I am thinking about writing a mathematical and scientific approach to the <u>Rule of St. Benedict</u> or towards the Benedictine way of life. Would you or Brother Gregory have any references or suggestions that might be germane to this task? If you want to include my phone number for the next Oblate meeting, it is 339-1748.

Yours in Christ,

Patrick A. O'Dougherty, Ph.D.

Bishop Lawrence Welsh: 11/19/96

Pat "Knocks the Dust off His Shoes":
Moving on From Newman Center

I have been at Newman Center on and off since the 1960s. The staff at Newman Center and my psyche counselors think that it is time for me to move on and to try some different intellectual irons in the fire. I published eight books while at Newman Center and built a research library in sustainable agriculture. Father Steve Bossi subtly picked a row with me. So, I threw plenty of cold water on Marxism plus God at Newman Center. I apologize for this. I admit that I raised issues that were offensive. However, what I am going to do is to dedicate my letters to the psyche dependent populations liquidated during WWII. There are cycles in Western Civilization; and, I don't want to see this one recur. Father Steve Bossi moved on. So has Archbishop Roach. So did Father Richard Colgan. I am going to follow suit. I've gotten some pretty good invitations.

I wish you well in all of your endeavors. I may go to the skyway blessing at St. Olaf's Thursday and show you my book on the Green Revolution. You are a distinguished priest and a prominent bishop.

God Bless,

Patrick A. O'Dougherty, Ph.D.

Bishop Lawrence Welsh: 12/6/96
The Definition Problem

IRISH PSYCHOLOGY/IRISH PSYCHIATRY: THE TIRESIAS COMPLEX

It was positive seeing you at the Tiny Tim Funeral. You are getting off to a good start as a bishop. I think you work well with artists and intellectuals. You need more definition.

I have finished a work on the Green Revolution called <u>St. Patrick, The Green Revolution, and the Hydrogen Conversion Project</u>. I am one of the main founders of the Green Revolution at the University of Minnesota. I hope to counterpoint Patricia Hampl at the University.
God Bless,

Patrick A. O'Dougherty, Ph.D.
See enclosures.

Sister Mary Anthony Wagner,   3/15/97
Oblate Crucible
I don't have a wife or children. I don't date. I think a life of carnality is a pitfall. My passions are writing and movements in history, psychology and religion. I have good luck with small groups of true believers. I left St. John's after my freshman year and joined a small group of radicals called Students for a Democratic Society. There were about eight of us in the group. Most of the group became famous. For example, see the letter I received from Vance Opperman who was in that group along with Lee Warren Smith, Howie Kaibel, Brian Coyle and me. Later, I ran for mayor of Minneapolis. We don't want to put up a Berlin Wall in the relationships in the Twin City Oblate group. Art Zannoni was in my Temple Israel group. Terry Gips was in the International Alliance for Sustainable Agriculture group. My mother is in a writers group. Linda Heyne got involved in our Oblate group in St. Paul and got a doctorate. Christ didn't want the <u>lukewarm</u> and neither do professors.

Irish Culture
I'm not German, however, I have taken different courses on German culture and have been a member of the Austrian Studies Program for 12 years. I have referenced a number of German thinkers, for example, Heidegger, Kant, Scheler, Nietzsche in my books. Maybe you don't know much about Irish culture. I am sending you an article by a man in my field, history, named Peter Berresford Ellis The article is <u>Revisionism in Irish Historical Writing: The New Anti-Nationalist School of Historians</u>. He is a mentor of mine. He has written fifty books. "Give it a try." **Cultures are "like apples and oranges."**
An emptying out,

IRISH PSYCHOLOGY/IRISH PSYCHIATRY: THE TIRESIAS COMPLEX

Patrick A. O'Dougherty, Ph.D.

S. Mary Anthony Wager
St. Benedict's Monastery
104 Chapel Lane
St. Joseph, MN 56374

Sister Mary Anthony,   4/25/97
Death Comes to the Msgr.
My uncle, Msgr. Loyola O'Dougherty, Vicar General, Tucson Arizona, died on April 9, 1997. I wonder if you could mention this in your Oblate letter. The liturgy at the wake was the Good Shepherd liturgy from St. John's Liturgical Press. The wake opened with Beethoven's "Hymn to Joy." A Latin band played at the wake. At the funeral the readings were from "The Book of Wisdom," the Eight Beatitudes, St. Paul's letter to the Philippians containing Christ's pouring out, a priest's prayer from the Book of Numbers. A Pima American Indian band played at the funeral. At the gravesite, "How Great Thou Art" was played as he was lowered into the grave.

Robert F. Kennedy Jr.
I had a Green Party meeting with Robert F. Kennedy Jr. on April 21, 1997. I talked with him about issues relating to the Communal Property Amendment. He autographed his flyer for me. I am sending you a copy of it. I sent those mail outs to Casey Alexander, Celeste Boda, Ann Condon, Susan Hulbert, Lorene Lawler, Ann Miller, and to you. I wanted to test the waters with the Oblate group. Casey Alexander, Celeste Boda, Ann Condon, Susan Hulbert and you showed up at the meeting in March. Clutch. Thanks for the feedback. I am going to send Robert F. Kennedy's autographed flyer to those four women. It might do something for their careers. I am going to wait a week before I do this. If you have a problem with this, call or write me. Then I am going to factor myself out of the equation.

Irish Psychology/Irish Psychiatry: Catholic, White Negro, Scientist, Revolutionist, The Tiresias Complex
I am almost finished writing the above book which addresses, analyzes, and organizes my letters. It includes the letters I have written the clergy and to Sister Mary Anthony. I have edited the letters and toned them down. I have also cleaned up the grammar. There were around eight hundred letters. I narrowed the number down. It is one of my most powerful works. I may get three published articles out the material that I sent the Oblates.

Up from slavery,

Patrick A. O'Dougherty, Ph.D., wanderer

IRISH PSYCHOLOGY/IRISH PSYCHIATRY: THE TIRESIAS COMPLEX

LETTERS TO COUNSELORS

To: Dr. Joseph Gendron
From: Dr. Patrick Strangethought

I have an interest in getting a non-traditional degree. There are several universities that offer them: "non-resident and short-residency Bachelor's, Master's, Doctorates and Law Degrees, also, Medical Degrees, Honorary Doctorates, Diploma Mills and much more." This quote is in <u>Bear's Guide to Non-Traditional College Degrees</u>. My dad got a J.D. degree when he was in his forties.

I want to make a movie--the anti-movie--no sex, no violence, "no plot, no character development." It is the pedestrian movie about America a nation of pedestrians. Sartre wrote some anti-novels; but, I think my movie would have more reality contact than his novels.

I'm just practicing on my computer, Joe.

Pat

4/5/94

Patrick O'Dougherty, Ph.D.
The Hellenist America Institute
Riverside Plaza M3410
1615 South Fourth Street
Minneapolis, Minnesota 55454

Joseph L. Gendron, M.D.
606-24th. Avenue South, Suite 515
Minneapolis, MN 55454

Dr. Gendron,

     I have been given the advice by Dr. Michael Franey of the Physics Department that I should not take on any new drugs right now. I should wait until there is a problem or a need and then take on the experimental research. He thinks that I should ask the doctors why I need a new drug **now**. Dr. Franey keeps in touch with me everyday. He does not think that I should have a drug prescribed for me based on how I was five or ten years ago. Dr. Franey was the person who initially told me to get psychiatric medications. He thinks that I am doing fine. Dr. Franey thinks that I should get a second opinion before engaging in this research. I think that I should take his advice. He got the research for me on my present medications. The drug manual he used said that the researchers did not understand how Prolixin worked. How is this new drug going to work? What am I doing now that would require it?

Patrick O'Dougherty, Ph.D.

cc. Dr. Abuzzahab Sr., MD, PhD.
Patricia O'Dougherty-Kast

Dr. Joseph Gendron,　　　　　6/16/94
Dr. F. S. Abuzzahab,

My mother, Patricia Kast, is coming to the meeting with Dr. Gendron on Monday, June 20, 1994. She and my sisters have concern about when I threatened intellectually Dr. Paul Murphy in the history department. My position is this. This professor is a pornography nut. He has spent most of his adult life championing the legalization of pornography and other <u>devious</u> first amendment issues. He helped sponsor a pornography show in the **chapel** at Macalaster College. He mixes up deviant sex with religion. I think that this guy is quite deviant. He doesn't have many emotions and compensates for this by devoting himself to deviant intellectual interests. I met another man in St. Paul who spent his time watching deviant sexual movies. I told him that I found this to be unacceptable. I have a degree in educational psychology and worked as a group therapist in the College of Education at the University of Minnesota. I know when to refer a client. I think that they hear at the history department that this professor is not completely normal. My uncle, John O'Dougherty and my friend, Mike Franey, think that this guy is deviant. This isn't an ethnic issue or a Catholic issue. It is a psychology problem.

I don't want to change my medications unless I have problems with the ones that I'm on.

Dr. Patrick A. O'Dougherty

cc. Patricia Kast, Margaret Wright, Mary Ann Alm, Maureen O'Dougherty
This has been a little hastily written.

June 7, 1994

J. Gendron, M.D.
606 24th. Ave. So.
Minneapolis, MN 55454

IRISH PSYCHOLOGY/IRISH PSYCHIATRY: THE TIRESIAS COMPLEX

re: F.S. Abuzzahab, Sr., MD.

Dr. Gendron,

Please accept this letter as a response to Dr. Abuzzahab's letter dated June 14, 1994.

He states that although the MMPI 2 of April 15, 1994 is completely normal, Mr. O'Dougherty is "exhibiting grandiose and paranoid delusional thoughts." Paranoia is a typical profile for writers and politicians. My mother, Patricia Kast, was with me at the meeting, and I am sure that she did not see any paranoid or delusional thinking.

Dr. Abuzzahab then makes two drug options. **I would like now to appeal to legal determination of my case in the courts and ask for a Medical Review Board to review my case.** I have been given the advice by Msgr. Loyola O'Dougherty to avoid Dr. Abuzzahab. Charlotte O'Dougherty thinks that I can do more for Dr. Abuzzahab than he can do for me. Dr. Abuzzahab is a fine man. However, I recommend Dr. Abuzzahab for Moral Therapy and Philosophical Therapy. It is not being paranoid or delusional to be a writer. I have written five books. I will write ten or twenty or thirty. I have taken the medical advice given me for years. I have been given the advice by a friend of mine Dr. Michael Franey to continue the drugs that I have been taking presently. I like my present schedule of drugs. I have had twenty years of daily suffering and I do not want any more. I have done nothing wrong. I am an outstanding patient. I want due legal and medical processes. I want to keep new drugs as an option. We have a law firm in our family.

Dr. Patrick A. O'Dougherty

cc. Patricia Kast, Margaret Wright, Mary Ann O'Dougherty, and Maureen O'Dougherty

July 7, 1994

F.S. Abuzzahab Sr., MD, PhD.
Clinical Psychopharmacology Consultants, P.A.
Ste. 303 Riverside Park Plaza
701 25th. Ave. So.
Minneapolis, Minnesota 55454

Dr. Abuzzahab:

IRISH PSYCHOLOGY/IRISH PSYCHIATRY: THE TIRESIAS COMPLEX

I am writing a response thanking you for your letter dated June 14, 1994. I am keeping it with my correspondence and I hope to get it in an archive someday. I have always had the same goal since I was a youth which is to write a classic. I went into intellectual history and the history of science to achieve this goal. I have chosen to write about many of the great thinkers: Kierkegaard, Sartre, Hegel, Thoreau, Shakespeare, for example. I do not think this is delusional thinking but normal behavior. I am tragically flawed. A psychiatric patient has not produced a classic and this would be a wonderful contribution to the field. I am sure there are many psychiatrists who have an interest in helping me accomplish this goal. I am going to keep your drug recommendations as an option. My friend, Dr. Michael Franey, thinks that I should have a legal case presenting my position before I have any medications changes. The Irish have produced many classics. I am writing my fifth book and I hope to write many more. I will let you know next summer how I am doing.

Patrick A. O'Dougherty, Ph.D.

I am a single man and I have my freedom. It is the single people who have made an awful many of the great intellectual and artistic accomplishments in the world.

Dr. Frank Preston,
Family Medical Clinics
3809 42 Ave. So.
Minneapolis, Minnesota 55406

6/13/95

Dr. Preston:

I filed these documents with several attorneys. I am a writer, intellectual, teacher, professor, activist, historian and many other vocations. I worked with Brian Coyle, Lee Warren Smith, and Vance Opperman in Students for a Democratic Society. I wanted to have some feedback and some interesting experiences to write about. So I got involved with a psychologist, Dr. Shirley Corrigan, back in the 1960s. I ended up getting processed with the courts and doctors. I had several Catholic doctors Dr. Patrick Stokes, Dr. Frederick Wilson, and Dr. Joseph Gendron. I always had a pretty normal personality profile on the scientific personality tests.I almost ended up dead with these doctors. Dr. Stokes told me that he couldn't figure out why I had chronic vomiting, diarrhea, sleeping problems and thoughts about death. He told <u>me to pick the pill or pills</u>. He did this with other clients too. Dr. Wilson charged me almost $900 for a few minutes visits in the hospital. He was later prosecuted for financial problems. With Dr. Stokes and Dr. Gendron, I had 19 years of daily suffering with tardive dyskinesia and other problems. I had major surgery due to a fall which I had in my apartment with the medications given me. My father, Aquinas O'Dougherty, almost ended up dead at St. Mary's Hospital in his forties because of a staphylococci infection that he contacted during a surgery there under the care of Dr. Charles Kelly. My father transferred to Mount Sinai Hospital from St. Mary's Hospital before he died. A nurse came into his room to insert a tube down into his intestines. It turned out that she had never done this procedure before. My uncle, Terence O'Dougherty, died at Fairview/St. Mary's after five major surgeries almost in a row. He got pancreatitis after one of the surgeries and died from it. A few of his last words were "get me out of the hospital." What would you suggest in my case?

Patrick A. O'Dougherty, Ph.D.

6/14/95

Dr. Joseph Gendron:
Medical Ethics and Activism
I want to look at the issue today of medical ethics and activism. This is a big problem in many Catholic countries and other countries as well. Intellectual and political courage and torture are the themes of <u>Kiss of the Spider Woman</u>. It is a Brazilian movie that I saw in Paris. Different people chose to do writing and activism in life, like Margaret Sanger, Gloria Steinem, Fathers Daniel and Philip Berrigan and Patrick O'Dougherty. How do you handle people like this? I have raised some big activist and money issues at Newman Center. My sister, Dr. Margaret Wright told me that they called you about this. Am I going to be busted? I have filed this situation with several attorneys. I have been fair enough with you about everything. Terence O'Dougherty thought, "You can get something on anybody." This is true for doctors also. You can always file something on a doctor. Medicine is a very imprecise science. I have a right to a private relationship with my doctor. Also, I have a lot of medical rights about drugs, for example. I have several professional friends. I can always get another doctor. We will see what happens.

Patrick A. O'Dougherty, Ph.D., activist

Joseph L. Gendron, M.D.
606 24th Avenue South, Suite 515
Minneapolis, MN 55454
6/19/95
Dr. Joseph Gendron:
Newman Center

IRISH PSYCHOLOGY/IRISH PSYCHIATRY: THE TIRESIAS COMPLEX

I am dropping contesting the budget issue at Newman Center. They are losing money. I have made my case.

Rights

I have a right to privacy. I have a legal right to refuse a drug. I have a right to medical access. I have a right to be a writer. I have a right to a doctor of my own religion or of my own choosing. I do not have to sign away my legal rights. Single people, like myself, are in the majority. They have rights. I have many other rights.

Request for a Medical Review and Legal Hearing

I almost ended up dead with Prolixin. I had major surgery with it, and I had tardive dyskinesia with many other adverse side effects like constant nausea, diarrhea, fainting and vomiting. I want to stay on my present prescriptions, and I do not want them changed without a medical review or legal hearing. How do we set them up? I am requesting them.

Patrick A. O'Dougherty, Ph.D.

Please place this letter in my file.

Dr. Joseph Gendron: 6/26/95

I got the prescription refilled that you gave me. Dr. Abuzzahab simply refilled this prescription and lowered the Benztropine Mesylate to 0.5 mg. I have the Geneva brand of Thorazine. I cannot possibly be an expert on drugs. Dr. Abuzzahab wanted me to sign away my legal rights. I'm not going to do that. Riverside Pharmacy wants me to get a new prescription. I want one ASAP. I can't be a single person here with the doctors. Most of my family doesn't even live in this state.

Newman Center/Issues/Intellectual Courage

The new Paulist Fathers at Newman Center came here from out of state. They want to give me no recognition, no concessions, and no coffee. Everybody thinks they are terrible to me. They have no dissent there other than me. The main issues relate to money and values. They have Catholic Marxism at Newman, that is, Brechtian liturgies. He was a Marxist dramatist. Terry Dosh, a church historian, laicized priest and Newman Center member, who taught me at St. John's University has a newsletter there, "Bread Rising." It looks up all the deviant and wayward priests and clergy in the world, for example, homosexual or unwed bishops. He is the head of CORPUS, the laicized-married priests' movement. I raised the Joycean issue of the queen bee and the drones with him. Joyce thought some of the orders in the Church were impressive, like the Jesuits; and some were drones, like the Christian Brothers. I don't agree with Joyce. People and religious orders evaluate on a case by case basis. Dosh reflects queen bee and the drone relationships. His wife supports him. Do you think the Church will go for these types of relationships?

IRISH PSYCHOLOGY/IRISH PSYCHIATRY: THE TIRESIAS COMPLEX

Newman Center is the pariah in the archdiocese. They want a real large piece of the money in the archdiocese. Most of the people are <u>not</u> for them in the archdiocese. I have been going to Newman Center on and off since the 1960s. I did not have a problem with Father Bury, Father Hunt or Father Alexander. I did not know Father Michael Joncas or Father Johnson. However, I heard that they were impressive. What would you suggest about this situation? Would you knuckle under?

Patrick A. O'Dougherty, Ph.

Dr. Joseph Gendron
606-24 Avenue South, Suite 515
Minneapolis, MN 55454

Dr. Gendron: 7/13/95

    My father, Aquinas O'Dougherty, had a law firm in Camden, Minneapolis. He and John Bohanon built an office building there back in the early 1970s. After his death in 1986, the office building was paid for. My father put the estate in my mother's name to avoid probate. My mother has not gotten one nickel out of that law office since my father's death. He turned over his practice to the lawyers there, John Bohanon and Jack Carleen, for <u>free</u>. He turned over the library in the law office there to these two lawyers for <u>free</u>. John Bohanon and Jack Carleen have been sitting on less than half the going rate of rent in that office building for <u>nine</u> years. The building is up for sale at a price of $160,000. They will never get that. The problem is that John Bohanon cannot afford to buy my mother out. Neither he nor Jack Carleen have much of a retirement income. They want to sit on almost free rent forever. Legally, my mother has rights to one half of each month's rent? This is the oldest game in the world--shakedown the widow. My mother is seventy-five years old. What do you do if two lawyers have you boxed in? My mother and Dick, my cousins and my brother can't take them on. I can get all kinds of free legal advice. What do you do with a situation like this? How do you force a settlement? John Bohanon is a Catholic. My mother and I are the only Catholics in our family. We want a deal. Values are the issue.

Patrick A. O'Dougherty, Ph.D.

cc Patricia and Richard Kast

Patrick A. O'Dougherty
Riverside Plaza M3410

IRISH PSYCHOLOGY/IRISH PSYCHIATRY: THE TIRESIAS COMPLEX

1615 South Fourth Street
Minneapolis, Minnesota 55454

Dr. Werner and Lee T.
Hennepin Faculty Associates
Adult Psychiatry
Hennepin Multispecialty Clinic
D110
900 South 8th Street
Minneapolis, Minnesota 55404

Dr. Werner and Lee T.
10/11/95
Regarding the Request for Medical Information

My internist Dr. Frank Preston, Family Medical Clinic South, recommended medication monitoring for me and a Catholic psychologist for my writing and counseling. Dr. Jean Jones, a psychologist, suggested I walk into the Hennepin Multispecialty Clinic because I live in that neighborhood. Medication monitoring would be convenient at that location. However, she does not have my medical records. I had an appointment with Dr. Frank Preston on 10/11/95 to discuss my situation. He suggested to me that I tell Dr. Werner that this is "a get acquainted interview to choose a psychiatrist." Dr. Preston suggested my medical records will transfer "if there will be continued care." His nurse, Kari, did not feel comfortable with the idea of releasing medical records to a doctor who I had never met. The privacy issue arose. I will show up for the 11-1-95 9 a.m. appointment with Dr. Werner and Lee T.

Patrick A. O'Dougherty

Dr. Frank Preston: **12/19/95**
To Your Retirement
A New School of Medical Thought: A Conceptual Framework
In honor of your upcoming retirement on Friday, December 22, 1995, I am founding the personalist/intuitionist school of internal medicine. Internal medicine is a personal/intuitionist point of view. A point of view is an activity. Intuitionism is the activity school of mathematics. Mathematical models of internal medicine's diseases, procedures, and drugs are personal activities. A **cutting edge** of research in internal medicine is finding alternatives to petro-chemicals which are activities. One of these alternatives will and its applications will prolong your retirement and your life.

IRISH PSYCHOLOGY/IRISH PSYCHIATRY: THE TIRESIAS COMPLEX

Yours in Christ,

Patrick A. O'Dougherty, Ph.D.

Dr. Joseph L. Gendron
606 24 Ave South Suite 515
Minneapolis, Minnesota 55454

Patrick A. O'Dougherty, Ph.D.
Riverside Plaza M3410
1615 South Fourth Street
Minneapolis, Minnesota 55454

5/1/96

Dr. Gendron:

Father Steven Bossi, Newman Center, a military man, a priest with major neurological problems and a major car accident

Five years ago they had a major staff change at Newman Center. Father Steven Bossi, an Italian Paulist priest, took the helm. He is a good enough man. I know him quite well because I go to daily Mass at Newman Center and have built a major agriculture research library in the Dorothy Day in the basement of Newman Center. He is a military man. It also turns out that he has some major neurological problems. He had a fall or blackout several years ago. Around the time of this incident, he had a major van accident on an Easter break trip to Appalachia. All the students could have been killed. He had to undergo a whole series of tests. I don't know the whole out come of these tests. He doesn't have the type of personality that can reach out to people very easily. For example, he cannot do male resonance very well. For all that I know, he might be experiencing a personal purgatory. Also, I did some research with Al Nier in the physics department. On the way to his home in St. Paul, something went out in his mind and he ran into a tree going thirty miles an hour and died. It was not a heart attack or a suicide. The deal is I know nothing about counseling priests, and I know little about neurology. You know a lot about counseling priests and neurology. Maybe this is why he called you?

At Loggerheads Situation with Father Steve Bossi

I am in the fields of intellectual history and the history of science. This kind of intrigued a military man and priest like Father Steve. He transferred to Newman Center from Clemson University in South Carolina. Anyway, Newman Center was not doing well financially. So, Father Steve drew a line with me over money issues and intellectual issues: no concessions, no recognition, and no rights. He took the

IRISH PSYCHOLOGY/IRISH PSYCHIATRY: THE TIRESIAS COMPLEX

initiative on this. I contested the Newman Center budget at the chancery to toughen the staff of <u>eleven</u> up. Most of the staff are very large people, two, three or maybe four hundred pounders. They loved me for doing this. This is why Father Steve called you up last summer about me in your office. I heard about this. I have a right to privacy and I value this in my relationship with all of my doctors. I want to feel comfortable with all of my doctors. This is a major issue in counseling. Father Steve is a confessor. He should know this. Anyway, I don't want to get involved in a no win situation at Newman Center when I have made a major contribution there is all dimensions. You've got a copy of my library. I also helped build the Lenore Scallen library there. You were a longtime member of the Newman Center congregation. So you know these are quite impressive contributions. I have also published seven books there. Several of them are in the areas of Catholic intellectual history. Nobody in the history of Newman Center ever did this. It is a formidable intellectual contribution on campus. The staff a Newman Center did not do this. I also presented the whole "white Negro" statement of the Irish there. During the "dark ages" of the 1950s there were signs <u>"Dogs, Irish and Negroes Do Not Enter This Pub"</u> throughout New England. They still have signs like this in English pubs. The Irish have a different trip than many of the congregation at Newman Center. The Irish can't build can't build an aircraft carrier. So, they have the grey and black realities to keep them clean. That is pretty tough stuff don't you think? Father Steve Bossi can't approach me on a personal level. I didn't know what to do about this situation. So I called the City Attorney and asked him what to do about it. He told me to file a report with the Minneapolis Police Teleserv. Anyway, I brought this up to the staff and Newman Center and had a mail out to some of the congregation about it. Father Steve didn't like this and neither did the staff. So, they told me they considered me to be a trespasser. Both Father Steve Bossi and I filed a report with the Campus Police about this. This stalemate over at Newman Center is wearing. I want the situation positively resolved without a whole lot of medical and personality problems interjected into it. Would he like it if I requested access to his personal medical health care givers? Many people do not know about his neurology problems and the van accident. A Catholic psychologist I know thinks his problem is that he is too Bossi-bossy. I used to be a counselor to young people and he wants to do this real definition with me and the young people. The deal is I do not want a staff of eleven to try to squeeze me out over there with no rights and no recognition. He is a strong leader, but is not good at the one to one. I am going to put the ball in his ball park and give him your phone number (612) 332-5577. You can bill my account at Minnesota Health Care Programs, Member Number 00173027. Or, better yet, why don't you bill him? I have raised all of these points with him. He is going to get a copy of this letter. I want to focus on my writing friends and Harvard.

Patrick A. O'Dougherty, Ph.D., the sparkplug at Newman Center,

IRISH PSYCHOLOGY/IRISH PSYCHIATRY: THE TIRESIAS COMPLEX

I think they should have a psychiatric neurology archive at Newman Center in the tradition of Joseph and Judy Gendron. Newman Center produced a famous French psychiatrist, Dr. Joseph Gendron. This should not be forgotten. I filed this situation with Father Mike Tegeder who happens to be enthusiastic about psychiatry and Dr. Joseph Gendron. <u>Dr. Joseph Gendron is the Dr. Franz Fanon of American psychiatry.</u> The whole field of psychiatry is basically "a pedagogy of the oppressed."

Please retain for my file.

cc. Father Steve Bossi and Father Michael Tegeder

Dr. Karen Bruggemeyer: 6/16/96
701 Park Avenue South
A Vietnam Veteran/An Irish Dissident/An Afro-American Attorney
Father Steve Bossi at Newman Center is a six foot Italian Vietnam Vet and a Southern facing military man. I'm a writer, dissident kind of a man--a formidable enough man. He drew a line with me about five years ago mainly over the money issue and intellectual issues. I've never had anybody do this with me before. However, I come from a big military family--twenty-five to thirty years active in WWII with five major campaigns. I know when this has been done. His liturgist must weigh close to four hundred pounds. I didn't start a fight over at Newman Center. I got along well with all the priests there. What would you do if some six foot male in the court systems drew a line with you? You are a specialist in forensic psychiatry. I did <u>not</u> take the initiative on this situation. <u>None</u> of my friends have been for this deal. I have several of them and many of them are prominent. I asked a black attorney at the Department of Human Rights what to do about this before I did anything. He mentioned the courts and the city attorney option came up. He also made other recommendations to me. I also filed this with a Catholic woman psychologist. Catholics have as many rights in the Church as patients do in American hospitals. I want to play this by ear. I thought you were a professional and competent doctor. I like the drug, Risperdal, so far. However, I just hope that it doesn't have a lot of adverse side effects. I was attentive to you as a physician.

Patrick A. O'Dougherty, Ph.D.

Dr. Karen Bruggemeyer: 6/21/96

To Albert Schweitzer: Publish or Perish
I thought that you might have an interest in a brochure of mine. The title is Beyond Vienna 1900: Rethinking Culture in Central Europe, 1867-1939. David F. Good, the former head of the Austrian Studies Program, invited me to do a book with

IRISH PSYCHOLOGY/IRISH PSYCHIATRY: THE TIRESIAS COMPLEX

him on "Reinventing Modern Austrian Mythology." I think what I'm going to do in response to my eleven year involvement in the Center for Austrian Studies is to write a work on Austria as Liturgy and Liturgy as Therapy. The therapy theme came as a response to a lecture I heard by Edward Timms, from the University of Sussex called "Group Dynamics in Vienna and London: The Psychoanalytic Society and the Bloomsbury Group." The women involved in the Austrian Studies Program are often the cream of the Germans. I taught in Graduate School. HCMC is a teaching hospital. The deal with graduate students is that if you don't have sparks and resolution they don't get an "A." What are your research projects at HCMC? If you don't break the ice with me you get a "B" from me as a doctor. I have involvement with Brother De Paul's mission to Haiti. This isn't some game going on here with Catholics. They have millions of people starving. I have been to Mexico and the Caribbean. They have to beg money for medical attention in those countries. Women have children in the streets in some of these countries. Catholics have millions of illiterate people. If you don't publish, they will perish. I want another trial or hearing in the government center for my response to the whole program here. My favorite psychiatrist is Franz Fanon, <u>The Wretched of the Earth</u>. He came out of the Caribbean and went to France. Neale Thompson told his interest in the chemical mapping of the brain. What are you interested in? People get jealous of Irish Catholics and envious of the Germans. Tell that to somebody who is starving or in desperate need of medical attention. I want to have an understanding.

Patrick A. O'Dougherty, Ph.D.

Dr. Karen Bruggemeyer, Dr. Robert Werner, Lee Tomsyck,
6/26/96
The Insecurity in Dealing with a Woman Problem
I am a single Irish Catholic male who has never married and seldom date. In certain situations and for certain issues I feel that I'm am on shaky ground dealing with a particular women. For example, I am an environmental ethicist. Also, I'm pro-life all species, for example, the tigers and pandas. I am for delimiting the use of military force. Some men do not like this position. St. Patrick did. I have a woman friend who is a vegetarian. I am not. However, I respect her position. Some Jews do not eat pork or favor cremation for burial. I do. I do not know what can be done about a problem like this. However, I want my position respected. You don't have to agree with it. Janis Vape was over at my apartment just the other day. He has heard about this. He knows that I'm taking my medications. Also, William Lubov told me that if I want to go over problems in medication monitoring to call him right away. This can be a big problem. I have a conscience, and if this is going to be a problem for any of you please call me. I don't want to make a big deal about this. Also, I agree that I do have a lot of accidental or tangential thinking. One

psychologist called it <u>creative</u> <u>divergent</u> <u>thinking</u>. However, when I was in the College of Education doing my student teaching at University High School Mr. Stockton told me that he gave me good marks on everything that I did except that I did not have that real professional teaching demeanor or persona. I do know if he used those exact words, however, I remember him telling me "I was the worst at that." I deny problems. Take my gambit for prizes with a sense of humor. "Humor is the best medicine." I want to raise a point about the medical students that I met at HCMC. I'm sure that they got me diagnosed the right way. I never intended to go to medical school. I don't like surgery or pathology, and I'm not mechanical so I wouldn't make a good surgeon. However, I'm a teacher and a writer and an intellectual. None of them shared any or their ideas or research proposals or the type of drugs that they were trying to get grants for or to get approved. They gave me an academic fright rather than a medical fright.

Patrick A. O'Dougherty, Ph.D.

Dr. Robert Werner,   7/26/96

I have been taking Risperdal since early in May. I have been having problems with excessive drowsiness, gagging, and a low sex drive since I have been on this drug. How do you address these side affects? I take one 3MG tablet twice daily. I may have also put on some weight. I have a concern about the drowsiness. Shalom,

Patrick A. O'Dougherty, Ph.D.

Joyce,

Franz Fanon
Fanz Fanon who wrote <u>The Wretched of the Earth</u> is my mentor in psychiatry. Have you read it? Psychiatry is a <u>Pedagogy of the Oppressed</u>, for example, oppressed women.

Dr. Robert Werner: 7/29/96
Defensiveness
The problem that I have on my MMPI is defensiveness and the doctors and nurses tell me there isn't a drug for it. What do you suggest when that is the case? For example, I'm not gay, but is there a drug for gayness? I am celibate. Is this a genetic chemical problem? Is there a drug for it?

Med Monitoring

IRISH PSYCHOLOGY/IRISH PSYCHIATRY: THE TIRESIAS COMPLEX

I ended up having in the past tardive dyskinesia, gastro-intestinal problems, and a sleeping disorder out of previous medications. This is why I am requesting some flexibility on medication monitoring. Could you give me a phone call about medication monitoring before my appointment with Lee Tomsyck on Thursday? The second Risperdal tablet on Thursday night put me out from 5:30 p.m. until 9:00 a.m. Friday. Joyce Forsgren suggested some reduction in dosage might be right.

On A New Field in Medicine
Did you like the essay that I wrote for you on alternatives to petrochemicals in medicine, for example psychiatry?

Intellectual Courage
I like to do topics in intellectual courage as well as academic preening. For example, I chose the Irish white Negro theme for intellectual definition. Do you have a problem with this? I ended up having some Viktor Frankl experiences because of this. How does one address experiences like these? Also, I would like you to return my book, ulster's white negroes when you are finished with it.

Shalom,

Patrick A. O'Dougherty, Ph.D.

Dr. Robert Werner
Dr. Karen Bruggemeyer
Lee Tomsyck, R.N.
Joyce Forsgren, Coordinator
Neale Thompson, Reverend
Janis Vape, Psychologist
Patricia and Richard Kast
Margaret, Mary Ann, and Maureen O'Dougherty

To the concerned: 9/11/96
Dr. Jane Rozsnafszky, Catholic Woman Psychologist, on Patrick O'Dougherty
I sent my HCMC file to Dr. Jane Rozsnafszky. She thinks I got "railroaded" on this hearing and stay of commitment, but it "worked out pretty well" for me. The power struggle at Newman Center ended with the transfer of Father Steve Bossi. Father Richard Colgan also transferred. She thinks my current dosage of Risperdal is working out for me. It is a good drug. **About my tests and personality structure, Dr. Rozsnafszky thinks I do not have a distinctly mentally ill personality.** I don't have

158

an abnormal personality on any of my tests. She doesn't pick up any abnormality in my relationship with her. She thinks I am very masculine, that I'm not particularly empathic and that I have a problem with defensiveness. She sees that I have no track record of violence. About the issue of threats, it is legal posturing. Father Steve Bossi, for example, called me up and <u>invited</u> me over to see him at his rectory. She doesn't see me as being particularly schizophrenic either or only mildly so. She thinks I have a lot of unusual thinking due to wide ranging reading which needs to firm up. Many people do not know how to take it. On my critical and negative orientation, she thinks it is a <u>family</u> personality characteristic. The family is critical, negative and achieving.

Recommendations

What she recommends for me is that (1) I continue in <u>Day Treatment</u> down at HCMC after my legal stay of commitment because it has been good for me and gives me a sense of validation, (2) <u>no more power struggles</u> because the one I am in is quite wearing on me, (3) for my unusual academic interests she recommends that I get on <u>Internet</u> because I won't find many people with my particular interests in bars maybe at HCMC or even at the university, and (4) she also recommends a <u>men's</u> spirituality group for me like the men's Oblate group at St. John's University. She also liked the idea of having me join the Green Party and become a kind of theoretician and revolutionist in it. Reverend Neale Thompson at HCMC liked this idea also. It is what I wanted to do in life. I may also work in the May Day Bookstore, an alternative bookstore, to carry out the aims of the Green Party. I want to continue building the IASA library. She also likes my level of spirituality though my sisters don't appreciate it. She thinks I should go back to Newman Center, keep a low profile, greet my friends there and just let the people accept me and see that I'm fine.

God Bless,

Patrick A. O'Dougherty, Ph.D.

Dr. Robert Werner,   8/13/96

Dr. Patrick Stokes who I had as a doctor from 1976 until the mid 1980s had me on the following drugs: Prolixin, Mellaril, Norpramin, Cogentin, and Benadryl. I was on most of these drugs all at once, and I don't recollect the exact dosages. The eye doctor, Dr. Shur, got him to end the Mellaril. I may not have spelled the names of all these drugs correctly. I felt that I was overmedicated with Dr. Stokes. I had problems with tardive dyskinesia, a sleeping disorder, and gastro-intestinal problems on these drugs. I was between sleep and waking much of the time and kind of experienced "the dark night of the soul" on these drugs. Dr. Stokes told me that he couldn't figure out the right medications for me. His strength as a doctor is that he has a buoyant attitude and gives good advice. You won't get your personal life messed up with him. After he went out of private practice to work for the Aspen Group, I switched to Dr. Joseph Gendron who my mother, Patricia O'Dougherty, worked with as a social worker. He gave me weekly injections of Prolixin, and changed me over to Thorazine. After this change, my sleeping problems went away. He also kept me on Cogentin for side affects. The problem with Dr. Gendron was that the tardive dyskinesia continued and I had a fall due to too high a dosage of Thorzaine. It was 100mg daily , but he reduced it to 25mg daily after the major surgery.

To try to remedy the tardive dyskinesia Dr. Gendron took me off the Prolixin about four years ago and the gastro-intestinal problems stopped. He sent me to Dr. F.S. Abuzzahab to see about a new drug experimental drug to replace Prolixin called Olanzapine. Dr. Abuzzahab wanted me sign away all of my legal rights and participate in a study with Olanzapine to see if it would work more adequately than Haloperidol. I declined to do this because of reasons of age. Also, I've had problems with other drugs and overmedication in the past. My friends didn't feel that Dr. Abuzzahab was right for a writer. Dr. Gendron wanted to do a writer and under his care I produced five books and built a major research library on campus. I had a lot of luck with the miracle drug, Thorazine. Both Dr. Stokes and Dr. Gendron are good doctors. I never missed an appointment in fourteen years with these doctors.

Risperdal

During my hospitalization in May, Dr. Karen Bruggemeyer dropped the Thorazine and the Cogentin because she thought my mind was "a little speeded up." She gave me a new drug, Risperdal. She gradually increased the dosage of this drug to 6mg daily. I feel the drug is a good drug. My mind is not speeded up as it was under Thorazine. I have noticed the following side affects under this dosage of Risperdal. First, I have had some concentration problems and it feels like there is tension on my brain front with the 6mg dosage of Risperdal. Second, I have had gagging and vomiting five times after meals. I eat my meals in restaurants so people think I'm an alcoholic with the gagging and vomiting. Third, I have had a problem of drowsiness at this level of dosage. Fourthly, I have very low sex drives under this dosage. I talked to Dr. Craig Qualey back in early June about what to do if you have a problem with vomiting with Risperdal. He told me to cut the dosage in half to 3mg. After I gagged and vomited twice last week, we discussed this situation and decided to vary the times when I took the drug. Dr. Jane Rozsnafszky, my consulting psychologist and Janis Vape, my Mental Health Case Manager, recommended going to Crisis Intervention if I have a problem with gagging and vomiting after meals again. To avoid doing this, I have cut the dosage of Risperdal to 3mg as Dr. Craig Qualey suggested I do if I have a problem with vomiting with

IRISH PSYCHOLOGY/IRISH PSYCHIATRY: THE TIRESIAS COMPLEX

Risperdal. I have noticed the following patterns happen under this dosage. First, I haven't as yet had a recurrence of the gagging and vomiting. I am keeping my fingers crossed. Second, the drowsiness has lessened. Third, my concentration is much better, and I don't feel a band on my forehead. The doctors told me if this is a problem please call immediately. The deal is I'm in the most productive years of my life. I don't want to be overmedicated or to go through hoops over a new drug or drugs. I have paid too be a price in life to do this. Dr. Karen Bruggemeyer is a good doctor. Her strength is that she is a consistent frame of reference and nondirective. The problem that I had with her is that I could not have one major conference with her about legal, Newman Center, family or academic issues. I think the approach to use with a writer is to let the reins out rather than pull them in. William Lubov, the attorney I had in the case that I requested in the Government Center, told me that if there was a problem in adjusting my medication to call him. Will this be necessary? I made some legal stipulations about the administration of Risperdal before I took the new drug. I want to go with the flow. I won the case at Newman Center and I just want to move on in life. I can give you all kinds of close professional references. I hope that I click with you.

Shalom,

Patrick A. O'Dougherty, Ph.D.

References
Dr. Mike Franey, physicist and confidant, 680-3792
David Noble, historian, 624-4190
Brother Gregory Conant, monk at Harvard, (508) 456-3221
Betty Agee, American Studies Department, 624-4190. She told me she doesn't want me overmedicated.
Sherrill Borkon of Temple Israel, 929-9520
Patricia Kast, my mother, (612) 743-2017 She used to work for Dr. Ted Liebermann, psychiatrist, at Hartford Mental Retreat.
Judith Rice, teacher/writer, 739-0926
Sister Mary Anthony Wagner, St. Benedict's Convent, (612) 363-7100
John Herlick, college roommate and friend, (218) 525-2699.
I can give you more.

Dr. Robert Werner,

10/9/96
Moving on from Newman Center

IRISH PSYCHOLOGY/IRISH PSYCHIATRY: THE TIRESIAS COMPLEX

On the letters that I sent you about Newman Center, I think it is time for me to move on and try some new irons in the fire. I am going to write them a positive formal letter. My medications have stabilized and I don't want them changed over this current deal at Newman Center. I don't want any more restrictions. I want to keep my options open. Also, I want to graduate from the HCMC Day Treatment Program on time on November 11, 1996. Janis Vape, my case manager, Joyce Forsgren my coordinator and Dr. Jane Rozsnafszky all think that I should knock the "dust off my shoes at Newman Center." Janis Vape who read the Newman Center letters I sent you, came over to see my on Monday. He thinks I'm doing great. I told him that I wanted to avoid future male ego contests. However, he told me that love and death and the conflicts in life are what makes it interesting. I saw Dr. Rozsnafszky on Tuesday. She thinks I have "stabilized" with my medication. I am addressing the personal validation issues with her. Many people have a personality structure similar to mine. Janis Vape thinks it is similar to the American one: critical, negative and achieving. How can I make it more accepting and positive? Americans, however, are quite psychopathic which I am not. To move on from Newman Center, I am working as a theoretician for the Green Party. I am trying to get Ralph Nader and Winona LaDuke elected president and vice president of the United States. I might write a book in the Green Party and put it on Internet. Sherrill Borkon, my friend from Temple Israel, sent me and invitation to a conference at Temple Israel sponsored by the Minneapolis Initiative Against Racism called Proclaiming Justice Throughout the Land. I told her that I was going to attend it. I am going to a three day symposium this weekend on Ludwig Wittgenstein at the Weisman Art Museum. The topic is Wittgenstein's Treatment of Knowledge in On Certainty. Wittgenstein is the philosopher who developed the field of the philosophy of language. This is an intellectual interest of mine. I am making some new contacts at the College of St. Thomas and at Cenacle Retreat House. My monk friend from Harvard, Brother Gregory Conant, OSB, was in town and I brought him to my analyst. He had never been to one before. He has pressured speech, religiosity, circumstantiality, intelligence and much good warmth. He feels his superior is a martinet. This is the wrong type of superior for a writer like Brother Gregory. He was a prisoner of conscience like I was. I brought him up to St. John's to present his book proposal to the Liturgical Press. They received it well. I am going to the psychoanalytic meeting at the YMCA this evening. My eighth book is at the University of Minnesota bindery. It is on the Green Revolution. The title is St. Patrick, The Green Revolution, and the Hydrogen Conversion Project. This book sparked my interest in the Green Party. I am soon writing my ninth book using a preface of Christian freedom to address my personal letters. I chose this preface because my Catholic countries do not have much freedom. My psychologist friend, Dr. Mark McGee, was in town for about three weeks and we went to the F. Scott Fitzgerald birthday party. Garrison Keillor autographed my program and Joseph

IRISH PSYCHOLOGY/IRISH PSYCHIATRY: THE TIRESIAS COMPLEX

Heller gave me his autograph. I, also, went to the Apollo Center twentieth year reunion and art show.

My sisters love me but they aren't particularly empathic towards me. How do you address that? They aren't all that different from me in personality structure either.

See you Monday, God Bless,

Patrick A. O'Dougherty, Ph.D.

For the HCMC file of Patrick A. O'Dougherty, Ph.D.
   A Summing Up of the Inpatient and Day Treatment Programs at HCMC
                              12/2/96
            The reason my family me <u>committed to HCMC</u> is to address <u>harassing</u> the staff at Newman Center and to address the problem of <u>tardive dyskinesia</u>. I have been over at Newman Center since the 1960s. In 1991 they had a change of staff at Newman Center of largely people from out of the state of Minnesota. For some reason they didn't appreciate me and subtly picked a row with me. Father Steve Bossi, an Italian, had <u>Catholic Marxism</u> or "Marxism plus God" for the agenda. For example, he had Brechtian liturgies, invited in Sandinista guerrilla leaders, Guatemalan insurgents, sponsored a fair play to Cuba approach, and had the Pathfinder Bookstore, a socialist bookstore, bring in its merchandise. Father Bossi also wanted to play the <u>race card</u> very hard. I do not oppose many different radical ideas. However, I decided after three years to contest this agenda on the basis that it isn't Catholic and that Newman Center should address a more universal agenda because it is underwritten by the archdiocese to the amount of $196,000 a year. The Church does not want politics in the Church. The staff at Newman Center did not appreciate my <u>contestation</u>. I ended up with no rights at Newman Center. I contacted a <u>black civil rights attorney</u> about this. He told me to contact the City Attorney's Office about it and do as the City Attorney instructed me. After Father Bossi heard about the legal posturing, a legal situation developed resulting in my commitment and <u>both priests leaving Newman Center</u>. I have decided to <u>move on</u> from Newman Center also. Joyce Forsgren, my coordinator, recommended this and Janis Vape my court psychologist <u>dismissed my case on 11/13/96</u>. I <u>tested normal</u> on all the tests given me at HCMC. The <u>tardive dyskinesia</u> problem has been addressed and the problem has diminished. Joyce Forsgren also recommended that I move on from Day Treatment. She contacted some of my professors through a friend of hers at the university. They told her that I was <u>brilliant</u>. I moved on after I received my certificate on <u>November 22, 1996,</u> the anniversary of the JFK assassination.The Inpatient and Day Treatment programs at HCMC impressed me.

IRISH PSYCHOLOGY/IRISH PSYCHIATRY: THE TIRESIAS COMPLEX

I am responding well to the <u>3mg dosage of Risperdal</u> prescribed for me by <u>Dr. Bruggemeyer and Dr. Werner</u>.

Patrick A. O'Dougherty, Ph.D.
FOR MY FILE IS A COPY OF THE <u>PUBLISHED BOOK</u> I PRODUCED OUT OF THIS EXPERIENCE ENTITLED, <u>ST. PATRICK, THE GREEN REVOLUTION, AND THE HYDROGEN CONVERSION PROJECT</u>. It is my <u>positive response</u> to <u>Catholic Marxism</u> at Newman Center. I am a major founder of the Green Revolution at the University of Minnesota and a member of the Green Party.

F.S. Abuzzahab Sr., MD, PhD.
Clinical Psychopharmacology Consultants, P.A.
Ste. 303 Riverside Park Plaza
701 25th. Ave. So.,
Minneapolis, Minnesota 55454

7/25/96

Dr. Abuzzahab:

I am sending you my personal resume and a list of my writing and research projects and the data base of the research library I have built for my file. I came into to see you with the <u>signs</u> of tardive dyskinesia problems due to the administration of Prolixin. The <u>symptoms</u> that I had reflect in the MMPI-2: The Minnesota Report: Adult Clinical System Interpretative Report which you administered to me on 15-APR. The enclosed <u>history</u> will bring my file up to date. I am presently a client of HCMC. I decided not to sign away my rights in the administration of Olanzapine because of my previous problems psychiatric medications. I am presently taking a small dosage of Risperdal.

Thanks for your time,

Patrick A. O'Dougherty, Ph.D. Irish white Negro, writer, activist.

Please retain enclosures for my file!

Jane Rozsnafszky, Ph.D.
Licensed Consulting Psychologist
2300 Humboldt Ave So.
Minneapolis, MN 55405

Dr. Rozsnafszky: 8/8/96

164

IRISH PSYCHOLOGY/IRISH PSYCHIATRY: THE TIRESIAS COMPLEX

I have been a psyche consumer since I was an undergraduate at the University of Minnesota back in the 1960s. In 1976 I got Dr. Patrick Stokes for a psychiatrist at Guild Hall. I had him for a doctor for about seven years. In about 1986 I got Dr. Joseph Gendron for a doctor after Dr. Patrick Stokes went out of private practice. Under the care of these two doctors, I got tardive dyskinesia due to treatment for schizophrenia. It is maximum daily discomfort which I had for about eleven or twelve years. After the discontinuation of Prolixin back in about 1992 and after major surgery due to a fall, the doctor placed me on Thorazine. The discomfort subsided and I have had a very productive life ever since. I raised the issue of tardive dyskinesia with Dr. Gendron and I think that he retired under fire last year. About one year from last June, I contacted the woman attorney at the Attorney General's Office who represents psyche consumers about the tardive dyskinesia I had. She recommended filing this against Dr. Stokes and Dr. Gendron's records. Do you think that I should do this?   Out of family and injustice issues at Newman Center, I requested a legal hearing in the Government Center and I agreed I would try a new drug Risperdal with the preconditions that the drug be withdrawn if there is a repeat of the tardive dyskinesia, gastro-intestinal problems or sleeping problems.I have had some gastro-intestinal problems which we have discussed. I talked to Dr. Craig Qualey about them in early June and he recommended the dosage of 3mg daily of Risperdal. I have been on 6mg daily. Would you have an interest in signing a release for my medical records?   I had Dr. Karen Bruggemeyer as an in patient and I have Dr. Robert Werner as an out patient.
A Dunce But Once,

Patrick A. O'Dougherty, Ph.D.

Dr. Robert Werner
Dr. Karen Bruggemeyer
Lee Tomsyck, R.N.

9/4/96
The Ecological Bases of Psychiatry
The Organic Based Psychiatric Drug Movement

I am going opposed to write a book in the area of psychiatry and found a movement in history in this field.   The book would be on the ecological bases of psychiatry. The movement in the history of the field of psychiatry would be The Organic Based Psychiatric Drug Movement. Think Organic, Feel Organic would be the slogan. However, this is not an easy project to do because I am not a chemist. Who are psychiatrists who know a lot about functional equivalents of chemical drugs, for example, organic based functional equivalents of Risperdal or Thorazine or Cogentin?

IRISH PSYCHOLOGY/IRISH PSYCHIATRY: THE TIRESIAS COMPLEX

Think Organic, Feel Organic,

Patrick A. O'Dougherty, Ph.D.

Dr. Joseph Gendron,

12/21/96

To Productivity

Season's Greetings! I liked Joseph Gendron as a doctor. He has a warm empathic personality. The sleeping problems I had got cleared up with the Thorazine. The Prolixin created tardive dyskinesia and gastro-intestinal problems. I had to address this is a tough minded fashion. However, the discontinuance of the Prolixin stopped the tardive dyskinesia and the gastro-intestinal problems. I haven't had daily suffering now since 1991. I am trying a Jewish doctor, Dr. Robert Werner. He has me on a small 3m dosage of Risperdal. This drug is effective.

I am moving on in life and in my relationships in life. I had several very productive years with Dr. Gendron.

God Bless,

Patrick A. O'Dougherty, Ph.D.

Dr. Robert Werner
Dr. Karen Bruggemeyer
Lee Tomsyck
Dr. Jane Rozsnafszky

1/3/97

The Contribution Productivity Approach

I went out to see Dr. Patrick Stokes who I had as a doctor for many years. He is an Irish Catholic like myself. I wanted a second opinion on the treatment program at HCMC; and, I had some real difficult medication problems with him which I wanted to address. Dr. Gendron figured out my medication problems. Dr. Bruggemeyer and Dr. Werner appear to have found a good drug, Risperdal, for me. However, Dr. Stokes does have a good approach. He focuses on how you can make a contribution. Contribution is the key to recovery. He focuses on productivity. Productivity is the key to recovery. He doesn't focus on how you can get all wrapped up in your own problems, or what is wrong with your family, or how you have a bad hand, or your largely negative diagnosis. He doesn't do a lot of theorizing. He wants you to focus on how you can get your life together quickly and make a contribution. This isn't an approach that Freud taught or Adler taught. I

166

IRISH PSYCHOLOGY/IRISH PSYCHIATRY: THE TIRESIAS COMPLEX

don't know if this an approach they teach in the psychiatry department or the psychology department. What do you think about it?

Patrick A. O'Dougherty, Ph.D.
cc. Dr. Patrick Stokes

Dr. Karen Bruggemeyer: 3/16/97
Diagnosis

I read over my file from HCMC. I don't disagree with your evaluation of me. However, I think I would like to make a qualification of this evaluation. I have a writer's and political activist's paranoid personality profile more than the schizoaffective-bipolar personality profile. I don't have mood swings. For example, I don't get depressed. However, I am socially outgoing and manic.

Family Problems

Ironically, I note in my file that Father Steve Bossi was the only one who put a good word in for me. He said I was quite an "unusual" man. I did give him several compliments, for example, for spiritual, intellectual, and leadership ability. Father Steve Bossi and his staff at Newman Center all published a lot of material with me. They made names for themselves on campus like I have. I come from a large Irish Catholic military family. Bill Caldwell, a counselor of mine, told me it reminded him of the book and movie, <u>The Great Santini</u>, by Patrick Conroy. He recommended I see this movie which I did. I thought there was a striking resemblance to my family in that movie. However, in contrast to the parents in the film my father and mother were both highly moral, spiritual, Catholic, political activists and writers like myself. <u>None</u> of my counselors have been much for my family. Joyce Forsgren told me to "find a new family." All of them want to be very tough and demanding on me in the military fashion for achievement. I am the oldest with the burden of status problem. They just reveal my sins, faults, negatives and weaknesses to people. They almost never give me any praise. For example, in my HCMC file Dr. Margaret Wright says that I said something to her son that she didn't like. <u>All I told him was that he was good looking and to be wary of strangers. We had a Jacob Wetterling case in Minnesota</u>. They have almost never been there for any of my successes, for example, my college graduation, my run for mayor of Minneapolis, my degrees, my promotions. I have many accomplishments. I have never been given credit for any of them. I made a lot of money for our family. I was in Massillon State Hospital in Ohio for <u>eight</u> months. <u>None</u> of my siblings wrote or called me. My mother and father were the only ones <u>loyal</u> to me. My brothers and sisters make me out to be a pariah. I feel I'm a true Christian in the family. I was one of the <u>loved</u> students in my high school. My brothers and sisters don't feel comfortable with me. I feel they reject and abuse me because of this. I'm try not to point up their faults. It would indict me.

Charity, The Mind, Catholic Spirituality, and Activism

I have devoted my life to charity. I will give you three examples. I have done Legion of Mary work for twelve years in nursing homes. I worked at Newman Center for six and a half years in the Dorothy Day room building a research library for free. I also worked for Little Brothers Friends of the Elderly. There are many. Would you do this?

On the mind, I'm a voracious reader and writer. I have always been that way. I have a immense thirst for knowledge. I have many published book and articles. I taught mathematics.

For Catholic spirituality, I have gone to daily Mass much of my life. I did an example of activism at HCMC. I got the clients at HCMC involved with my Green Party campaign and produced a major published work out of it called, St. Patrick, The Green Revolution and The Hydrogen Conversion Project.

I have also taken many risks in life. This is a plus.
A Social Justice Issue
I had a social justice issue in my life. I worked daily at Newman Center for six and one half years without compensation. They had a change of staff at Newman Center, and the staff was jealous and mean spirited with me. I went out of my way to resolve the situation. It was not resolvable. I had to stand up for myself. I had to seek some kind of legal remedy. Instead of getting praise or seeking justice in this situation, none of my family put a good word in for me, my brothers and sisters were not loyal to me, none of them were accepting of me the way I am, none of my brothers and sisters were compassionate about me. My sister, Peg, poisoned the well with her children for me during the situation. Earlier in life, she told me she had a high paranoid score on the MMPI. My sisters are South Carolina Irish Catholic Northern girls. I got dealt a grey card here. They let me take a beating on it. I cried for two days about this. Mary Ann showed the downside of her personality to me as she always does with everybody. Maureen remained aloof, reserved and argumentative as she always is. My mother is loyal and compassionate about me and cries about me when I have a problem. I am the one most like her in the family. However, she doesn't have any brothers and doesn't feel comfortable with me. When I was about four years old, we were living over at 4104 Bryant, I had a disagreement or fight with a neighbor boy, Bobby Truax. She remembers that incident as the day that I went wrong in life. She has found out what is wrong with me practically everyday since. This was twenty-one years before the onset of

schizophrenia. By focusing on this all the time, it is just her way of emphasizing what is wrong with me. My father's family has always been there for me and liked me and encouraged me. They don't focus on what is wrong with me. Instead, they think I'm intelligent, moral, interesting, a good conversationalist, a good writer, a good teacher and first rate. In contrast, my mother and sisters do not feel comfortable with me. I feel they are merciless with me about this. They don't accept me or <u>forgive me</u> or encourage me, or praise me. I feel forced into a corner with them. However, I used to beg my father to tell my mother to quit nagging me. He never did. When I brought my father to Guild Hall, the residents responded well to him. This is true when I talked about him at HCMC. When I talk about my mother and sisters, I am given this type of advice. "Don't internalize it." "Move on." "Shut the door on the relationship." "Get a new family." However, I'm a single person in life. Who will I have then? The big problem in <u>The Great Santini</u> is between the son and the military father. In contrast, in our family it is between the son and the mother and daughters. My sister, Maureen, doesn't do this. However, my sister, Peg, and my mother tell other people my sins, weaknesses, faults and shortcomings and then they tell me about them. My sister, Peg, has seldom been there for one of my successes. Here is an example of how paranoid my sister, Peg, is. My sister, Mary Ann, is a musician and has invited me to many of her concerts. I was in Ohio visiting my sister, Peg, and I had never heard her give a lecture. I thought it might be an intellectual treat to hear one of her lectures. So, I asked her if I could sit in on one of her lectures or classes. She could tell her students I was her brother visiting from Minneapolis. I thought she would take my invitation as a compliment. She told me she did not want me to do this. She <u>did not feel comfortable with it</u>. It is not appropriate. Some of the students might not like having someone who is not a member of their class be there. I have many sins, weaknesses, faults and shortcomings. My mother and sisters don't try to have understanding or reconciliation or just accept me. My mother, my sisters, and my brother's wives all have very strong dominating personalities. What do you do about a problem like that? My father in contrast was a very humble, intelligent man. What do you do about a problem like that. I cry about it. I pray about it. I seek counseling about it.

Consider the Sources

When people read my file, I would like to give them some advice. There is practically nothing positive about me in it. I feel the medical classification is also largely negative. My family has placed a lot of negative content in there about me. <u>None of my friends or associates put anything negative in there about me</u>. Instead of focusing on my sins, faults, and shortcomings. Please consider the sources. My case is before God.

"My Weakness, My Strength"

People think I'm intelligent. However, what they focus on with me is that I'm unusual or different, or schizophrenic. My high scale on the MMPI is K. The raw score is 22 which is nearly two standard deviations high. In comparison my paranoid raw score is 10 and my schizophrenia raw score is 9 which are just around one standard deviation high. To address the defensiveness, I maintain that I'm quite flawed. To address the paranoid schizophrenia, my confession is "my weakness is my strength."

With prayers and love,

Patrick A. O'Dougherty, Ph.D.

POSITIVE NOTE FOR MY COUNSELORS

One of the main criticisms made of Patrick O'Dougherty by his counselors is that he is critical, negative and makes intellectual threats. I had some tough things to say about my counselors in my letters because of my problems with tardive dyskinesia. I am placing a good word in this book for Dr. Patrick Stokes and Dr. Joseph Gendron. Both doctors have very normal personalities, and they always gave me a lot of encouragement. They both gave me very good personal advice. They are both good Catholic doctors also.

LEGAL LETTERS

John and Jack,   10/4/94

I got a Ph.D. and I have published five books and two articles. I want to publish around twenty books. I have an interest in starting a Catholic Studies Department at the University of Minnesota. I want to get a priest on the faculty of the University of Minnesota. They have never had a priest on their faculty. I hope that we get an honest settlement out of the law office. Our apartment building and condo sales went well.

I would like to see you again sometime.

Patrick O'Dougherty, Ph.D.

Patrick A. O'Dougherty, Ph.D.
Riverside Plaza M3410
1615 South 4th. Street
Minneapolis, Minnesota 55454

IRISH PSYCHOLOGY/IRISH PSYCHIATRY: THE TIRESIAS COMPLEX

Eric Cooperstein
Office of Lawyers Professional Responsibility
25 Constitution Avenue
Suite 105
St. Paul, Minnesota 55155-1500

1/15/95

Eric Cooperstein:
Assistant Director

NOTICE OF COMPLAINANT'S RIGHT TO APPEAL

In the Matter of the Complaint of Patrick A. O'Dougherty against John C. Bohanon and In the Matter of the Complaint of Patrick A. O'Dougherty against John C. Carleen dated January 2, 1996, the Complainant has <u>tentatively</u> decided to take the consul of Kenneth L. Jorgensen and Eric Cooperstein. They suggested "the more appropriate forum for complainant to resolve this civil dispute is the district court system." My father, Aquinas O'Dougherty, deceased attorney whose estate is in question, thought it is a wise idea to take attorneys' advice if offered. However, unlike doctors advice, he usually took his own legal advice. Thus <u>the notice</u> of complainant's right to appeal fourteen days after the January 2, 1996, like the estate of his father is <u>held in abeyance on today's date January 15, 1995.</u>

Dr. Patrick A. O'Dougherty and The School of Critical Legal Studies

The school of Critical Legal Studies tries to critique the <u>ideology</u> and <u>practice</u> of all aspects of the American legal system. This school adapts "a de facto ethnographic approach to legal education, the spoken and written discourse of legal professionals, and the social effects of how the system operates in practice as opposed to formal models, to which legal scholarship, closely allied with legal practice, has been prone to, but also to show how law as a process operates contrary to conventional wisdom." I appealed to the Office of Lawyers Professional Responsibility for justice in the name of my father's legacy and estate held in abeyance. Eric Cooperstein, Assistant Director at the Office of Lawyers Professional Responsibility, suggested keeping a dialogue about this legacy and estate. The question I would like to raise is does the Office of Lawyers Professional Responsibility <u>operate</u> "contrary to conventional wisdom." I mentioned to Eric Cooperstein the basis of Plato's, <u>Republic,</u> an ethnographic dialogue, is justice. This is a new interpretation. Isn't this the basis of the Office of Lawyers Professional Responsibility and of our American government? Critical Legal Studies tries to critique cultural hegemony, and authoritative meanings and processes. Are these problems in this complaint and in the Office of Lawyers of Professional Responsibility? Reference: George E. Marcus and Michael M. J. Fischer, <u>Anthropology as Cultural Critique: AN EXPERIMENTAL MOMENT IN THE HUMAN SCIENCES</u> (Chicago: The University of Chicago Press, 1986), 154.

171

IRISH PSYCHOLOGY/IRISH PSYCHIATRY: THE TIRESIAS COMPLEX

To Be Published, Please Respond,

Patrick A. O'Dougherty, Ph.D.

Note: Eric Cooperstein did not sign the determinations. Kenneth L. Jorgensen did. He was later referred to me by the Office of Lawyers Professional Responsibility.

Patrick A. O'Dougherty, Ph.D.
Riverside Plaza M3410
1615 South 4th. Street
Minneapolis, Minnesota 55454

Kenneth L. Jorgensen
Office of Lawyers Professional Responsibility
25 Constitution Avenue
Suite 105
St. Paul, Minnesota 55155-1500

1/15/95

Kenneth L. Jorgensen:
First Assistant Director

NOTICE OF COMPLAINANT'S RIGHT TO APPEAL

In the Matter of the Complaint of Patrick A. O'Dougherty against John C. Bohanon and In the Matter of the Complaint of Patrick A. O'Dougherty against John C. Carleen dated January 2, 1996, the Complainant has <u>tentatively</u> decided to take the consul of Kenneth L. Jorgensen and Eric Cooperstein. They suggested "the more appropriate forum for complainant to resolve this civil dispute is the district court system." My father, Aquinas O'Dougherty, deceased attorney whose estate is in question, thought it is a wise idea to take attorneys' advice if offered. However, unlike doctors advice, he usually took his own legal advice. Thus <u>the notice</u> of complainant's right to appeal fourteen days after the January 2, 1996, like the estate of his father is <u>held in abeyance on today's date January 15, 1995.</u>

Dr. Patrick A. O'Dougherty and The School of Critical Legal Studies

The school of Critical Legal Studies tries to critique the <u>ideology</u> and <u>practice</u> of all aspects of the American legal system. This school adapts "a de facto ethnographic approach to legal education, the spoken and written discourse of legal professionals, and the social effects of how the system operates in practice as opposed to formal models, to which legal scholarship, closely allied with legal practice, has been prone to, but also to show how law as a process operates contrary to conventional wisdom." I appealed to the Office of Lawyers Professional Responsibility for justice in the name of my father's legacy and estate held in

172

IRISH PSYCHOLOGY/IRISH PSYCHIATRY: THE TIRESIAS COMPLEX

abeyance. Eric Cooperstein, Assistant Director at the Office of Lawyers Professional Responsibility, suggested keeping a dialogue about this legacy and estate. The question I would like to raise is does the Office of Lawyers Professional Responsibility operate "contrary to conventional wisdom." I mentioned to Eric Cooperstein the basis of Plato's, Republic, an ethnographic dialogue, is justice. This is a new interpretation. Isn't this the basis of the Office of Lawyers Professional Responsibility and of our American government? Critical Legal Studies attempts to critique cultural hegemony, and authoritative meanings and processes. Are these problems in this complaint and in the Office of Lawyers of Professional Responsibility? Reference: George E. Marcus and Michael M. J. Fischer, Anthropology as Cultural Critique: AN EXPERIMENTAL MOMENT IN THE HUMAN SCIENCES (Chicago: The University of Chicago Press, 1986), 154.

To Be Published, Please Respond,

Patrick A. O'Dougherty, Ph.D.

Patrick A. O'Dougherty, Ph.D.
Riverside Plaza M3410
1615 So. 4th. Street
Minneapolis, Minnesota 55454

Kathy Kosnoff, Attorney at Law
Minnesota Disability Law Center
430 First Avenue North, Suite 300
Minneapolis, MN 55401-1780

1/17/95

Kathy S. Kosnoff, Attorney:
Private and Confidential

This is a response to your letter dated January 9, 1996. I am a writer, an activist, and a psyche consumer. I got a psychologist, named Dr. Shirley Corrigan, while I was an undergraduate at the University of Minnesota. I continued this course until the present day. The deal is if you are a writer you can't just learn from textbooks or experts. You have to amplify experiences. Also, you need to have an indeterminate position in society for literary freedom. Instead of seeking a corporate job, I walked out of the door in graduate school with very little money. I hitchhiked down to Atlanta, Georgia, to see what would happen. A lot happened. Anyway I ended up getting processed by the courts and the social agencies. I have always had a normal profile on the MMPI. I ended up getting tardive dyskinesia from

173

IRISH PSYCHOLOGY/IRISH PSYCHIATRY: THE TIRESIAS COMPLEX

Prolixin. I had this problem for eleven or more years. Dr. Patrick Stokes and Dr. Joseph Gendron were responsible for this drug prescription. I suffered horribly from this drug with psychic dependence, tremors, pseudoparkinsonism, falling, inability to get deep sleep, and almost daily vomiting. Under Dr. Joseph Gendron, I ended up having major surgery on my knee from falling three years ago. He removed the drug and put me on Thorazine. This drug enabled me to sleep. However, I feel there is a social control issue here with Prolixin. I raised the issue of major surgery with Dr. Gendron and the tardive dyskinesia contraindications. He decided to end his outpatient practice and semiretired rather than deal the legal problems posed by this suffering. He gave me some psychiatric referrals. What are my legal and patient rights in a situation like this? What are my rights in the area of literary freedom as a writer and psyche consumer? Could you delineate them for me? Kate Millett was another local writer and psyche consumer. She addressed her problems in a book, The Loony-Bin Trip. I've had enough suffering. Could you send me the leaflet on complaint procedures which you recommended? How do you rate my intellectual, religious, social, and political contributions coming out of a psyche consumer experience?

A pedagogy for the psyche resurrected,

Patrick A. O'Dougherty, Ph.D.
See the enclosed list of some of my accomplishments and writing projects.

Patrick A. O'Dougherty, Ph.D.
Riverside Plaza M3410
1615 South Fourth Street
Minneapolis, Minnesota 55454

Kenneth L. Jorgensen
First Assistant Director
Office of Lawyers Professional Responsibility
25 Constitution Avenue, Suite 105
St. Paul, MN 55155-1500

1/24/95

Mr. Jorgensen:

Legal Accountability
Buckpassing in the Government? A Shell Game?

I received a Determination dated January 2, 1996, and signed by you In the Matter of the Complaint of Patrick A. O'Dougherty against John C. Bohanon and a Determination in the Matter of the Complaint of Patrick A. O'Dougherty against John C. Carleen. I called over to your office to talk to you about this inquiry. They told me another attorney is handling this Determination. I talked to him and he told me that a Member of Board, another attorney, is going to rule on the appeal. Why didn't you respond to my inquiries? I made a response in a postmarked letter dated January 15, 1996. This response is within the appeal deadline stated in the Determination. **In fairness to me**, I think this is like buckpassing or a shell game, or "the run around" in the Office of Lawyers Professional Responsibility. I want an **abeyance** on this Determination until I have a clear idea who I am dealing with in that office. I don't want to have somebody claim Kenneth L. Jorgensen signed this Determination. You should have talked or dealt with him. Then I received a referral and another referral and a final determination. I expect **accountability** for **this** Determination by the person who signed the form, Kenneth L. Jorgensen. Isn't that correct legal procedure? I was given the advice at your office to keep the dialogue on this matter open.

Please respond for **your** Determination in writing,

Patrick A. O'Dougherty, Ph.D.

Margaret A. Russell
Managing Attorney
Legal Aid Society of Minneapolis
430 Avenue North, Suite 300
Minneapolis, MN 55401-1780

Margaret A. Russell: 7/13/95

My father, Aquinas O'Dougherty, had a law firm in Camden, Minneapolis. He and John Bohanon built an office building there back in the early 1970s. After his death in 1986, the office building was paid for. My father put the estate in my mother's name to avoid probate. My mother has not gotten one nickel out of that law office since my father's death. He turned over his practice to the lawyers there, John Bohanon and Jack Carleen, for free. He turned over the library in the law office there to these two lawyers for free. John Bohanon and Jack Carleen have been sitting on less than half the going rate of rent in that office building for nine years. The building is up for sale at a price of $160,000. They will never get that. The problem is that John Bohanon cannot afford to buy my mother out. Neither he nor Jack Carleen have much of a retirement income. They want to sit on the office building with almost free rent.   Legally, my mother has rights to one half of each month's rent? This is the oldest game in the world--shakedown the widow. My mother in seventy-five years old. What do you do if two lawyers have you over a barrel? My mother and Dick, my cousins and my brother can't take them on. The lawyers are just sitting on the estate. What do you do with a situation like this?

IRISH PSYCHOLOGY/IRISH PSYCHIATRY: THE TIRESIAS COMPLEX

How do you force a settlement? John Bohanon is a Catholic. My mother and I are the only Catholics in our family. We want a deal. Values are an issue. I want to focus on spirituality and writing after my fathers's death.

Patrick A. O'Dougherty, Ph.D., client
Please retain for my records.

Complainant's Name:
Patrick A. O'Dougherty, Ph.D., heir
Riverside Plaza M3410
1615 South Fourth Street
Minneapolis, Minnesota 55454
Telephone: Home (612) 339-1748 Work: (612) 339-1748

Office of Lawyers Professional Responsibility
Lawyers Professional Responsibility Board
25 Constitution Avenue, Suite 105
St. Paul, MN 55155-1500
(612) 296-3952

                                    12/8/95
Name of Lawyer: John C. Bohanon
Address: 4149 Lyndale Ave. North
City, State, Zip Code Minneapolis, Minnesota 55412
Telephone: (612) 529-9647

Name of Lawyer: John C. Carleen
Address: 4149 Lyndale Ave. North
City, State, Zip Code Minneapolis, Minnesota 55412
Telephone: (612) 529-9647

Complaint:
Professional Responsibility Problems: Major Neglect and Delay and Money and Accounting Problems

"In the Name of the Father"
Major Neglect and Delay Problems
Nine Years And Not A Penny On A Deceased Attorney's Estate!

My father was Aquinas James O'Dougherty, a deceased attorney in the State of Minnesota, who passed away **October 5, 1986**. He practiced law in the Claims Department of Northwestern National Life Insurance for John S. Pillsbury, Jr., for ten

176

years. He joined the law firm of Haughland, Carleen, O'Dougherty and Bohanon. Carleen and Bohanon are the remaining members. **Since my father's death in 1986, our family has not received a just settlement out of his estate.** Aquinas O'Dougherty turned his law practice over to John C. Bohanon and John C. Carleen. He also turned the law library of this law practice over to them. This law practice and law library have probably been quite lucrative for both of them. **Our family never received any money for this law practice or this law library. Before his death, he had to rewrite his partnership with John C. Bohanon and John C. Carleen because they were getting expensive cuts on <u>his</u> settlements.**

Money and Accounting Problems

My father had a very friendly, outgoing personality. He was somewhat naive about people including John C. Bohanon and John C. Carleen. I found the buyer for the apartment building that my father and mother owned. However, my father built an office building with John C. Bohanon. John C. Carleen could not afford to join in the ownership of this law office/business located at 4149 Lyndale Ave North, Minneapolis. This building has been paid off for several years. **Our family has not gotten one penny out of this twenty-five year investment.** The rent in the building is less than half the going rate of rent in the city. John C. Bohanon claims he does not have enough money to buy my mother's interest out. The building is on the market for an **intentionally** high price of $160,000. He could easily get a mortgage on this new property. I contacted a business real estate agent at Edina Realty about this. **Because the building is free of debt, he told me that my mother has rights to half of each month's rent. She has not gotten this for several years. <u>This is one of several probable major money and accounting problems</u>. How much rent is paid by John C. Bohanon and John C. Carleen pay for their law office in this building? They are probably paying quite little compared to comparable other pieces of property.** My mother, Patricia Kast, was Hubert Humphrey's Vice-Chairwoman for the Hennepin County DFL. She is seventy-five years old. She has moved to Florida. Her sister, who is not much older, died recently. These are the oldest games in the world--sit on the estate and bilk the widow and children.

"Judgment Day"

Darren at the Attorney General's Office and Anne at the Office of Lawyers Professional Responsibility suggested filing a complaint about this situation. Our family does not want to be over a barrel with two lawyers if my mother, who is not a lawyer, should die. I would like a full, just and ethical investigation of this situation. I will be willing to get the <u>Minnesota Rules of Professional Conduct of the Minnesota Statutes</u>, Vol. 52, as was suggested to me by your office, to review all possible points of professional misconduct in this situation.
Patrick A. O'Dougherty, Ph.D., heir

IRISH PSYCHOLOGY/IRISH PSYCHIATRY: THE TIRESIAS COMPLEX

Attention: Anne

John Bohanon and John Carleen:

2/2/96
Fathers and Sons
"You Reap What You Sow"
"The Chickens Will Come Home To Roost"
Plains' Truths

I have an interest in resolving this legal situation amicably. A famous story in my father's life was after the death of his father during the Great Depression when he was thirteen. A Maple Lake, Minnesota, farmer hired him to work on his farm. He worked all summer for this farmer doing all the hard farm labor, for example, the sowing and threshing. At the end of the summer the farmer went to Aquinas' mother, Emma. He paid her <u>thirty-seven dollars</u> for my father's labor for the whole summer. This was practically slave labor wages. The irony of this story is that three weeks later this farmer had an accident. He hit his head in the kitchen and died. He met his maker on that note. I wonder what his eternal reaping was? Later on in life, Aquinas and his brother went to work for some German nuns on a farm in Colorado building a barn for them. The movie, <u>Lilies of the Field</u>, arose out of this story. It starred a black man in the role of my father. Both of these stories are plains' truths stories of the Great Depression. Aquinas' chickens are coming home to roost. You wonder which story will bear the grains of truth here. My father was a classicist. Like Plato's, <u>Dialogues</u>, I am going to leave the ending of this dialogue inconclusive and open.

Patrick A. O'Dougherty, Ph.D., Catholic Dissident.

Patrick A. O'Dougherty, Ph.D.
Riverside Plaza M3410
1615 South Fourth Street
Minneapolis, Minnesota 55454

Marcia A. Johnson, Director
Office of Lawyers Professional Responsibility
Minnesota Judicial Center
25 Constitution Ave., Suite 105
St. Paul, MN 55155-1500

Marcia A. Johnson, Director,   2/8/96

"Top of the Pyramid"
Sloppy Professionalism

IRISH PSYCHOLOGY/IRISH PSYCHIATRY: THE TIRESIAS COMPLEX

I would like a complete review and written response by you of the complaint and appeal I filed with your office against the attorneys, John Bohanon and John Carleen. First, I want to address the issue of professionalism. Kenneth L. Jorgensen of your office signed the Determination dated, January 2, 1996. I tried to reach him on the phone within the fourteen day duration of the appeal deadline. He was unintentionally never available. He referred the case to Eric T. Cooperstein who told me that he was going to have a Member of the Board decide on the appeal. I filed a postmarked response within the allotted period and raised the issue of an illegal shell game or buckpassing in the government over the appeal of this Determination. I requested in writing an abeyance on the appeal of the Determination to deal with the problem of buckpassing in this Determination by Kenneth L. Jorgensen. Eric T. Cooperstein <u>agreed</u> to the abeyance over the phone. When I contacted Kenneth L. Jorgensen in writing about dealing with the illegal buckpassing and maneuvering, he told me he would not allow an appeal because I did not file within the allotted time period. I did. There is some <u>lying or carelessness with the truth here</u>. He did not give a legal reason for this action. Kenneth L. Jorgensen and Eric T. Cooperstein are backing down from a justified investigation of the complaints I filed. Are they having a good scare? Are they in over their heads in that office?

Anti-Catholicism

Second, I am addressing is the issue of Catholicism and Catholic ethics. I am an Irish Catholic writer who came out of St. Paul and Minneapolis. I have many prominent friends in both cities. I feel discrimination is an issue here. For example, I raised the approach of Critical Legal Studies as my framework of analysis for this Determination's appeal. This is an ethnographic cultural critique of the American legal system. My treatment by the staff of your office, discrimination, validates this Critical Legal Studies school of thought. Will I have to take further action to deal with this discrimination?

"Top of the Pyramid"

Third, I sought advice about the Determination made by Eric T. Cooperstein and Kenneth L. Jorgensen from a prominent St. Paul Catholic politician. He told me you are dealing with the middle of the pyramid, Kenneth L. Jorgensen and Eric T. Cooperstein, at the Office of Lawyers of Professional Responsibility. Go to the top of the pyramid. You are the Director of that office. I am appealing to you for a complete legal review in writing of this whole situation. Also, who is the Head of the Board of the Office of Lawyers of Professional Responsibility?

Top of the Alps,

Patrick A. O'Dougherty, Ph.D.

Patricia M. Siebert, Attorney

IRISH PSYCHOLOGY/IRISH PSYCHIATRY: THE TIRESIAS COMPLEX

Minnesota Disability Law Center
430 First Avenue North, Suite 300
Minneapolis, MN 55401-1780

Dear Patricia M. Siebert: 2/13/96
This a letter written in response to your letter dated February 12, 1996.

Thank you for giving me the advice about my mother being the beneficiary and "the other owner of the building in question, she would be the one with the standing to raise any issues related to the income from the building, and probably also any assets from his law practice." I <u>will</u> take this advice. I have filed my problems with tardive dyskinesia with several doctors and with three unnamed Catholic women psychologists. I ended up with eleven or more years of unnecessary "maximum daily discomfort" from this problem. I had major arthroscopic surgery from a fall in my apartment because of this problem. I want legal forms signed and cosigned by a lawyer or lawyers before any further medication changes. I talked to a lawyer at the Known Your Rights Law Firm, Schwaebel, Goetz and Sieben. I have a right to do this. Do lawyers in your office do this? I have not been hospitalized for eight years. I have always worked. I want all the proper legal, patients rights, signed and filed before any future hospitalization. I have never had EST. I will <u>never</u> give my consent to have it. My doses of neuroleptic medications are low. I wish to keep them that way. <u>I take my medications religiously</u>. Medicine is a very imprecise science. It is more of art than a science. A goal of the Hippocratic Oath is the alleviation suffering. I have had enough. I will keep the advance psychiatric directive as an option. I want legal advice about it if I do fill it out. I have an unnamed psychiatrist, GP, and consulting psychologist to help me make medical and mental health care decisions. Therefore, I will keep the proxy option open. I am forty-nine years of age and have hopefully many productive years left. I have many friends and family members who can intervene if I am ever "viewed as lacking the capacity temporarily to make those decisions." I will keep the proposed HEALTH CARE POWER OF ATTORNEY form in my archive. After my death, I wish all of my medical, psychiatric, and psychology records released and placed in my archive. I am part of the anatomy bequest program at the University of Minnesota. This program provides a free cremation after the use of the cadaver by the anatomy department. I wish to have my remains interred in an inexpensive crypt at Lakewood Cemetery. It is permissible for Catholics to be cremated as long as their remains are interred.

Psychiatry: Professional Contributions

I have invented a new field in psychiatry and psychology called personalist intuitionism. Personalism is a school of thought favored by Catholic thinkers, like

180

IRISH PSYCHOLOGY/IRISH PSYCHIATRY: THE TIRESIAS COMPLEX

Dorothy Day and Karol Wojtyla. This school emphasizes the rights of the person in all his or her dimensions, for example, spiritual, intellectual, patient, social, and legal. Intuitionism is the activity school of mathematics. Mathematics is an activity, like the mind, advocacy, birth, patients rights activism, or the creation of the universe. This school of mathematics stands in opposition to Plato's ideal forms. In law, I would apply this philosophy to the school of Critical Legal Studies which I favor.

The Alternatives to Petrochemicals Project

I am now doing research on alternatives to petrochemicals in medicine and many other areas. One example is the use of alternatives to petrochemical based drugs in psychiatry. The field of alternatives to petrochemicals will revolutionize most areas medicine, especially psychiatry.

The Harvard Benedictines

I have received a Harvard nod and I want to focus on that invitation at this stage of my life. I am sending you a brief list of my accomplishments. Please keep these materials for your file.

God Bless,

Patrick A. O'Dougherty, Ph.D

February 14, 1996

Patrick A. O'Dougherty, Ph.D.
Riverside Plaza M3410
1615 So. 4th. Street
Minneapolis, Minnesota 55454

Mr. Gregory M. Bistram
Chair of the Lawyers Professional Responsibility Board
Office of Lawyers Professional Responsibility
Minnesota Judicial Center
25 Constitution Avenue, Suite 105
St. Paul, Minnesota 55155-1500

2/14/96

Mr. Gregory M. Bistram:

IRISH PSYCHOLOGY/IRISH PSYCHIATRY: THE TIRESIAS COMPLEX

I am writing to you to request a complete review of the Complaint against John C. Bohanon and John C. Carleen. I would like a written response of this review.

Pharaoh's Game: Steal the Rents

John C. Bohanon and John C. Carleen are engaging in the oldest game in the world: steal the rents. Since my mother's office building in Camden has been paid off, we have rights to one half of each month's rent. This rent amount grows from the time it has been paid off until the present with interest. My mother hasn't received anything during this period. The period has been almost three years. John Bohanon, the co-owner of the building, has no right to steal the rents each month and claim it for a manager's fee, or claim business is tough or deal himself low rents. This is all specified in law. Dr. Mike Franey, John Harris who is a Harvard Law friend of mine, an Edina Business Realtor, and Patti Sullivan, attorney, told me this. We want a check out of John Bohanon every month until the property sells. Money is the key to a good relationship. If John Bohanon wants to remain in that property, he has to pay us every month for our half interest in that property. John Bohanon has only the right to claim documented costs. We are willing to allow for that. I appealed to the Office of Lawyers of Professional Responsibility to investigate this situation. They are in over their heads on this case. How do you resolve a situation like this one? Should we go to criminal court or civil court in this matter?

Patrick A. O'Dougherty, Ph.D.

Patrick A. O'Dougherty, Ph.D.
Riverside Plaza M3410
1615 South Fourth Street
Minneapolis, MN 55454

Gregory M. Bistram
Lawyers Professional Responsibility Board
Minnesota Judicial Center
25 Constitution Avenue
Suite 105
St. Paul, Minnesota 55155-1500
Mr. Bistram: 3/9/96

The Verdict

The life and death of my Father, Aquinas O'Dougherty, on October 5, 1986, coincides with the symbolic death of the American legal system.

About your letter dated March 6, 1996, I make the following observations about the legal profession: First, three or four years ago the Vatican condemned the American legal profession as being "corrupted and morally bankrupt." Second, the

sheer numbers of lawyers in this country have created the largest "internal brain drain" of any civilization in the history of the human race. For example, there are about as many lawyers in the State of Minnesota as there are in the country of Japan. Third, lawyers as a group economically just "cut the pie" rather than make it or bake it.[193] Fourth, the life and death of my father, Aquinas O'Dougherty, on October 5, 1986, coincides with the symbolic death of the American legal system. The verdict is unanimous.
Patrick A. O'Dougherty, Ph.D.

Patrick A. O'Dougherty, Ph.D.
Riverside Plaza M3410
1615 South Fourth Street
Minneapolis, MN 55454

Marcia A. Johnson
Lawyers Professional Responsibility Board
Minnesota Judicial Center
25 Constitution Avenue
Suite 105
St. Paul, Minnesota 55155-1500
Marcia Johnson: 4/1/96
The Verdict
The life and death of my Father, Aquinas O'Dougherty, on October 5, 1986, coincides with the symbolic death of the American legal system.
In response to the letter by Gregory Bistram dated March 6, 1996, and the other correspondence on the complaints against attorneys John Bohanon and John Carleen, I make the following observations about the legal profession: First, three or four years ago the Vatican condemned the American legal profession as being "corrupted and morally bankrupt." Second, legal ethics have become and oxymoron. Third, the sheer numbers of lawyers in this country have created the largest "internal brain drain" of any civilization in the history of the human race. For example, there are about as many lawyers in the State of Minnesota as there are in the country of Japan. Fourth, lawyers as a group economically just "cut the pie" rather than make it or bake it.[194] Fifth, as a single person which makes me a member of the new American majority, I think I have been discriminated against by your office. Sixth, I raise a question about the power structure of the Lawyers Professional Responsibility Board. Is it queen bee and drone relationships? Seventh, the life and death of my father, Aquinas O'Dougherty, on October 5, 1986, coincides with the symbolic death of the American legal system. The verdict is unanimous.
Please file this response with my complaints against John Bohanon and John Carleen.

IRISH PSYCHOLOGY/IRISH PSYCHIATRY: THE TIRESIAS COMPLEX

Patrick A. O'Dougherty, Ph.D.
cc. Kenneth L. Jorgensen, Eric Cooperstein, and Gregory Bistram

UPDATE
**THE LAW OFFICE DEAL WAS SETTLED AMICABLY ON MARCH 31, 1997
AFTER <u>ELEVEN</u> YEARS FOR NEARLY $50,000. I SENT JOHN BOHANON A THANK
YOU CARD FROM ST. PATRICK'S GUILD FOR RESOLVING THIS SITUATION
BEFORE MY MOTHER'S DEATH. I ALSO SENT HIM AN AUTOGRAPHED FLYER
SIGNED BY ROBERT F. KENNEDY, JR., ATTORNEY, WHO I HAD A MEETING WITH
IN DENVER AS A WAY OF THANKING HIM.**

LETTERS TO DAVID NOBLE
Dave,   5/23/94

  I think there should be a Catholic synthesis on campus. I think there should
be a priest or religious on the faculty of the University of Minnesota. I think that this
country needs another revolution or a Cromwell. I have a complex personality and
I am not going to harass any professors. My Dad was a drill instructor in the military
for a while and he could handle any problem that was thrown his way. He could
handle sex, or psychic, or racial problems or Ivy League professors. Can you handle
a guy like Andrew Greeley, or Avery Dulles or Malcolm X? I'm going to stay focused
on writing and spirituality and the Catholic synthesis. You don't have to take my
advice, but why not have a Catholic synthesis on campus. The door should open
here. It would be the beginning of a big contribution. I've got limitations to my
personality, but why not focus on my strengths. I'm sending Jackie Onassis a
remembrance card.

Brother Patrick O'Dougherty, Ph.D.

Friends in the History Department,   5/19/94

  I want to found a Department of Catholic Studies at the University of
Minnesota. I would also like to refound the University of Minnesota, America and
Western Civilization like St. Stephen, St. Paul, Daedalus, or Socrates did. I am
modeling my life after Orestes Brownson. I was a socialist in the 1960s and I want
to finish life as a Radical Catholic Republican like he did. I am going to try to publish
twenty or more books. It is time to get this department off auto pilot. If you share
similar interests to mine, call me at the Hellenist America Institute/O'Dougherty
Foundation at 339-1748.

IRISH PSYCHOLOGY/IRISH PSYCHIATRY: THE TIRESIAS COMPLEX

Dr. Patrick Aquinas O'Dougherty

David,   9/29/94

    I got a big article published in the Minnesota Daily on Newman Center. This young student intern, Chris Fellerhoff, at Newman wanted to test himself out on me-- a Young Turk. He gave a sermon at a Mass sponsored by Father Rich Colgan a priest at Newman. Chris structured the sermon around the phrase, "God damn." He used this expression many times in the sermon. I wrote a letter to the <u>Daily</u> opposing this approach. I launched two Catholic Germans right. I shouldn't try to make it for another guy. I will call you later on in the quarter. They have a Catholic studies department at Princeton, don't they?

To Royalism,

Dr. Pat

                            10/29/94
David,   **1/2**
    Nature is a royalist/aristocratic principle. Americans have lost one half.

    Royalist: "an adherent of royalism; person who supports a monarch or a monarchy, esp. in times of revolution, civil war, etc. specif., a) a supporter of Charles I of England; Cavalier b) a supporter of the British in the American Revolution; Tory c) a supporter of the Bourbons in France."[195]

    Tory: "1 a) in the 17th cent. any of the dispossessed Irish who became outlaws, killed English settlers and soldiers, and lived by plundering b) later, an armed Irish Catholic or Royalist 2. in 1679-1680, a person who opposed the exclusion of James, Duke of York, from succession to the English throne 3. after 1689, a member of one of two major political parties of England: opposed to Whig, and later to Liberal, Radical, Laborite; changed officially c. 1830 to Conservative 4. in the American Revolution, a person who advocated or actively supported continued allegiance to Great Britain 5. any extreme conservative; reactionary--adj. of, being, or having the conservative principles of a Tory."[196]

    Radical: "b) favoring fundamental or extreme change; specif., favoring basic change in the social or economic structure."[197]

    I like Royalism/Toryism and I joined the radical/socialist movement on campus: double trouble, double bind.

IRISH PSYCHOLOGY/IRISH PSYCHIATRY: THE TIRESIAS COMPLEX

Royalism and radicalism are <u>both</u> major themes in Catholic history--the Hapsburgs and Dorothy Day are examples.

I think that you are doing an excellent job.

Patrick O'Dougherty, Ph.D.

Memories of Charles Dickens' <u>Bleak House</u>
David,   1/8/95
The Government Light of Day: Project Come Clean
I don't know if you have ever had any business deals. We have several of them in our family: an apartment building, a sold condo, a sold house in Minneapolis, a law office building, a home up at Clearwater, a home in Winter Haven, Florida, a lot at Voyageur Village. We have come out on all of them. However, we have never gotten a penny out of my father's old law partners and the office building he built with one of them. I asked at the Attorney General's Office about this. They referred me to the Office of Lawyers Professional Responsibility. Two of the women there recommended filing a complaint about this situation. I filed it. Kenneth L. Jorgensen, First Assistant Director, filed the reply. I called over there to talk to him. He never responded. The staff referred me to another attorney, Eric Cooperstein, to deal with this case. He didn't sign the form. He told me if I appealed this case he would refer it to an unnamed board member who would rule on it. I have two weeks to appeal this case. I will appeal it. After this ruling, there would be no further appeals unless the facts change. Will the facts ever be investigated in this case? The laws are often easy to determine but the facts aren't. Both of my Catholic doctors decided to retire. I have two Jewish doctors. I pray that I have luck with them. Cooperstein may be Jewish also. I think with lawyers it is a good idea to throw the situation into our American government. I won't mention your name to these lawyers. However, I want them to come clean. I filed this case with a consulting psychologist and a Mental Health Advocate. I am waiting for the dust to settle on this one. I am going to pay this office a visit.

Patrick A. O'Dougherty, Ph.D.
See enclosures!

David,   1/25/95
"In The Name of the Father"
Aquinas O'Dougherty, My Father, Catholic Scholar, War Lover

IRISH PSYCHOLOGY/IRISH PSYCHIATRY: THE TIRESIAS COMPLEX

My father's family had twenty-thirty years active in the military with five major campaigns. My dad went from scratch up through the ranks and was one of the top of his class at OCS. He became a drill instructor at OCS after graduation. What type of male do they pick for that type of job? When I was a junior in high school I gave him Andre Malraux's, <u>Man's Fate</u>. I told him this is what I wanted to do in life. He never had a recruit like that but he was for it. When I was a senior he was working at Northwestern National Life Insurance and involved in undercover intelligence surveillance of local radical groups with John Guize. I asked him the question, "What type of man goes to work for another man?" He quit his job and went into private practice. I didn't feel comfortable with the war in Vietnam because he served on the Pacific Front during WWII. So I got involved with many of these radical groups and opted for psychological disobedience right after Robert Kennedy's assassination. The psychologist I sought advice from, Dr. Shirley Corrigan, was a military nurse. She agreed with my assessment about a probable loss of the war in Vietnam. Dr. Corrigan told me to have the doctors at the draft board call her. She told them to forget it. She spent years in the military working with military doctors. My dad liked Thomas Wolfe as a Southern writer, especially <u>You Can't Go Home Again</u>. When I was in graduate school he wanted to have me pulled out of there and toughened up as much as possible. I opted for the psyche trip and the Deep South. He let me get eight months in Atlanta, Martin Luther King's city of emergence, on Bellwood Chain Gang in the hospital section. The sentence was for a very minor dash and dine offense. One of my arrests was right by his church in Atlanta. Later Patrick moved on to Harvard, specifically the Harvard and Radcliffe Benedictines. There is a lot of literature coming out on how the Cold War has shifted to America. I am going to win it. A lot of European countries like Intellectual History, for example, prisons, psyche evaluations, philosophy, and diplomacy. Back the Green, the Grey and the Purple Viconian Revolutions, Patrick A. O'Dougherty, Ph.D.

Let the Legion of Mary forgive and sanctify this revolution through space and time.

Dear Friends in the History Department:
4/4/95
Patrick O'Dougherty's Project: Refounding World Culture

I brought up the idea of refounding or reinventing world culture at all levels to David Noble, Mike Franey, Steve Bossi, Terry Dosh, Mark McGee, F.S. Abuzzahab during the June of 1994.
This letter serves as copyright protection of this idea.

Patrick A. O'Dougherty, Ph.D.

IRISH PSYCHOLOGY/IRISH PSYCHIATRY: THE TIRESIAS COMPLEX

4/23/95

Dear Friends in the History Department:

THE SCIENCE OF PUBLIC ADMINISTRATION SHOULD RUN THE HISTORY DEPARTMENT AT THE UNIVERSITY OF MINNESOTA

I think they should make a science of student selection, graduate school selection, professor hiring, promotions, and professor productivity at the history department at the University of Minnesota. There at least should be a scientific weighing in the selection and promotion decisions. They shouldn't just have a personality merry-go-round there. History is not exactly a science, but many universities try to make public administration a science. Give it a try. I have a relation who was the administrator of a university in the Minnesota school system, and he thinks half of the professors are deadwood. I think a judge should hear what is going on here.

St. John's University: Academic Freedom

When I was at St. John's University, a distinguished professor from the University of Minnesota gave a talk to our class. He told us there was more academic freedom at St. John's than there was at the University of Minnesota. Another professor from the University of Minnesota told us that the honors papers at St. John's were as high as level of quality as most of the Ph.D.s produced at the University of Minnesota.

A Writer's Personality

I have a writer's personality. I have always had a **normal** personality profile on scientific personality tests. I do real well on the aptitude tests. I wanted to be a writer, instead of an attorney like my father, Aquinas, an attorney for John Pillsbury. You have to experience events to write about them. You just can't learn to be a writer from textbooks. I am going to try some affirmations over at the history department.

"Cultural Counter Revolutionary,"

Patrick A. O'Dougherty, Ph.D.

David Noble,

11/15/95

The Teaching Profession and the Soul Issue

188

IRISH PSYCHOLOGY/IRISH PSYCHIATRY: THE TIRESIAS COMPLEX

Much of the teaching profession in America has lost its soul. In Europe thinkers, like Edith Stein, traditionally regarded teaching as a spiritual profession. The reason so many immigrants are coming into America from the third world, especially Africa, is that America has lost much of its soul. Immigration is a "soul train."

T. S. Eliot and David Noble: "incarnation"

The central idea of T.S. Eliot in the relationship of religion and culture is "incarnation" rather than the words "relation" or "identity." A culture, he thinks, is the "incarnation of its religion." The word "incarnation" changes the whole meaning of the Old Testament. How do you relate the word, "incarnation," to the words eternal, Adam, T.S Eliot, New World and Garden, nature, literature, royalism, progressive historian, time, "transcendence of time," covenant and innocence? These are themes you like.

Is David Noble a theologian of history?

Patrick A. O'Dougherty, Ph.D.

David Noble,   11/19/95
An Antiuniversity?
I want to drop the history department personality issue. However, on a campus you can have evil triumph over wisdom. If the material triumphs over the spiritual, humanity can sink back into the goo. The East Germans and the Eastern Europeans tried a socialist experiment with atheism. I just feel Americans, for example, at the University of Minnesota, are repeating many of their intellectual, political, social, and personal mistakes. For example, Middle East literature began with the embrace of God in the Bible. Antiliterature and "antihistory" break that embrace. Marxist literature is an antiliterature and "antihistory." Brother Gregory Conant, O.S.B., maintains a Pope started the first university in Western Civilization. Has the University of Minnesota become an antiuniversity?

Secular Triumphalism
I don't really feel there is a whole lot of dissent on campus. There is more at Boston College. **The major intellectual premise at the University of Minnesota is to counter the Catholic and Christian synthesis. Then, they tolerate few Christians on the faculty and almost no Catholics to oppose secular triumphalism.**
Happy Thanksgiving,

Patrick O'Dougherty, Ph.D

IRISH PSYCHOLOGY/IRISH PSYCHIATRY: THE TIRESIAS COMPLEX

I got an invitation and couldn't edit this letter until Sunday. The Irish have a fight and the black and grey realities to keep them coming up roses, right?

Dr. John Howe,   11/26/95
Catholic Studies Dept./Hydrogen Conversion Project/Personalist-Intuitionist School of Philosophy/IASA

I want to start a Catholic Studies Department at the University of Minnesota. I want to get a priest on the faculty at the University of Minnesota. For example, Father Stransky has inaugurated the 1995 Burke Lectures on Religion and Society at the University of California, San Diego. It is a ten year old program. Why can't they have a similar program at the University of Minnesota? They want this at Harvard. I am also enclosing a copy of my Hydrogen Conversion Project. It is part of an alternatives to petro-chemicals project that I am founding. I have built an internationally recognized sustainable agriculture library on campus. It could go to Yale. I have also started the personalism-intuitionist school of philosophy and physics. This whole school of activism and personalism-intuitionism is a powerful framework for analysis and religious and scientific synthesis.
See you Thursday,

Patrick A. O'Dougherty, Ph.D.

Dedicated to the 1 billion people who are starving to death in the world.

David,   1/25/96
Beyond Einstein
"Albert Einstein was Intellectually Lazy." Aquinas O'Dougherty

The hydrogen conversion project, especially from sea water which accounts for nearly 70% of the earth's water, the alternatives to petrochemicals project, for example, medical research alternatives and the application of the intuitionist school of mathematics to these problems are three of the most critical research initiatives in the history of science, and the history of the University of Minnesota and in the history of military science. Many small countries in the world like Ireland, Israel, Latvia and Nigeria have large coastlines. These issues are life or death survival issues for them. Russia is a large landlocked country. America, in contrast, has a large vulnerable coastline. The most critical military issue for this coastline is to convert the seawater there to hydrogen. This is true for Africa and Australia as well. For example, Australia does not have a large supply of petrochemicals. Neither does Japan. The basis of activism is the activity school of mathematics.

IRISH PSYCHOLOGY/IRISH PSYCHIATRY: THE TIRESIAS COMPLEX

Patrick Aquinas O'Dougherty, Ph.D., Activist, The Green, the Grey and the Purple Viconian Revolutions.

An Activist Personifesto

The Germans have many strengths. However, they, as people, don't go against existing condition much. They almost never had a revolution in their history. Until recently, they, as a people, weren't real big on activism. For example, Petra Kelly, a founder of the Green Party in Germany and an activist, died in a murder/suicide. This just happened. The exceptions prove the rule. I am going to chose activism, the activity school of mathematics, and street knowledge for my strengths.

David Noble, Wanderer,   1/26/96
Patrick O'Dougherty and the Kissinger Nomination

I worked on in graduate school in the history department during the war in Vietnam the Henry Kissinger appointment. Very few people thought he would get it. He was born in a foreign country. He was Jewish in a yet untested civil rights movement. The appointment was to the state department. It is almost impossible to get a foreign national in a country's state department, especially during wartime. My mother was Hubert Humphrey's Vice DFL Chairwoman of Hennepin County. I knew Humphrey and worked for him since I was in grade school. I got published in grade school in an essay or letter on Catholicism and civil rights. Humphrey got me a job at the post office when I was in high school. I wrote several letters to Humphrey about having a Jew involved in diplomacy during a war against Marxists. Kissinger is that Jew. Humphrey approved the appointment I think largely to my pleadings. I thought the Jew should be a Republican because diplomacy is the forte of the Republican Party.

Patrick O'Dougherty: Redefining the Wanderer: God in The Statistics of Life

As a result of this effort, the construct I chose to define myself intellectually is that of the **intellectual wanderer** in all dimensions, for example, spiritually, politically, and psychologically. Did you work on that nomination? Who did on this campus. I don't want to claim credit for <u>any</u> of his later accomplishments.

Amnesty International

I told Sister Mary Anthony Wagner at St. Benedict's College about my minor prison sentences in the South. She has spent years working for Amnesty International. They should get a Nobel Prize also.

PATRICK O'DOUGHERTY, PH.D., WANDERER, THE GREEN, THE GREY, THE PURPLE REVOLUTIONS

David Noble,   **David Noble and Dante**

191

IRISH PSYCHOLOGY/IRISH PSYCHIATRY: THE TIRESIAS COMPLEX

2/8/96

David Noble argues most of our major novelists from James Fenimore Cooper to Saul Bellow are philosophers and theologians who repeatedly test the national faith in an American Adam living in a New World Eden against innocence as the American situation. The national community defines itself as a congregation of earthly saints living outward from historical culture. The novelists have little choice but to make their heroes philosophers and theologians who contradict their initiation into the realities of American with this ideal. Cooper, Hawthorne and Melville negate the idea America can become a New World Eden. They reject the heavenly city on earth as a defensible ideal. They refuse to believe in the perfectibility of the person. America has become a Garden of Gethsemane rather than a Garden of Eden.

Dante: Hell as a Starting Point

In contrast, Dante's starting point in the Divine Comedy, his autobiography, is Hell. The theme of the Divine Comedy is "purgation to purification." Dante rises through the rings of hell upward and finally ascends with Beatrice on a rose to the Divine. His call is to perfectibility of the individual and to the transcendence of love in the universe. **David Noble's challenge as an intellectual is to heed Dante's call.**

Patrick O'Dougherty is an American Catholic intellectual who chooses wandering as a methodology and "purgation to purification" to define himself to echo Dante who moved from hell as a starting point to the perfectibility of the individual and the triumph of transcendent/metaphysical love in the universe.

See David Noble, The Eternal Adam and the New World Garden: The Central Myth in the American Novel Since 1830.
Patrick A. O'Dougherty, Ph.D. a wanderer upward from the purgatory part of Minneapolis and St. Paul the starting point.

Dave, The Vatican condemned the American legal system. Georgetown University has dropped Chaucer, Milton and Shakespeare from their curriculum.

David,   2/13/96
The Science of Theology
The Dialectical Relationship Between Faith And Reason
Catholic theology is a science. This is the basis of the thought of St. Thomas Aquinas. The idea is to spiritually and scientifically integrate life. The problem many Europeans have with Protestantism is that it is not coherent and individualistic. There are twenty-five thousand Protestant sects. Is this rational? Many Protestant sects do not have a scientific theology. Is fundamentalism scientific. The idea behind Martin Luther is a "dialectic between faith and reason." Is this the basis of

your thought? I heard the University of Minnesota is not going to have a religious studies program. Are faith and dialectics taboo at the university?

Catholic Dissent?

I play a dissenting card intellectually in America without being too offensive about it. Maybe I should be offensive. What do you think about this? Priests, ministers and rabbis should sanctify through space and time this green, grey and purple American revolution. I have a pretty strong suit on this campus intellectually. I like the group in the American studies program. I feel comfortable with them.

The business deal got sorted out.

Happy Valentine's Day,

Patrick A. O'Dougherty, Ph.D.

David Noble: 3/6/96
The problem is Hyman Berman and John Howe both have these fragile male egos. It is their weakness and strength.

Irony of Ironies

Hy Berman: more Christian than many Catholic clergy

The irony is Hy Berman, a Marxist and intellectual Jew, was more of a Christian to me than most of the Catholic clergy I've dealt with including Archbishop John Roach. I caught hepatitis while I was in graduate school while I was taking his seminar. There is no cure for it, and it is a potentially fatal disease. I called Hy up from the hospital and told him that I might have to drop out or postpone work in his seminar. He asked me if I become his T.A. I have done many activities over the years with him. For example, I visited him while I was an undergraduate when he was teaching at the University of California at Berkeley. After dealing with Hy, I can see why Christ was a Jew rather than an Irishman. Hy was more charitable to me than many Christians. I give him a good review.

Patrick A. O'Dougherty, Ph.D., Catholic

David Noble: Archcritic of Capitalism
Resolving the Paradoxes and Contradictions of Capitalism

Senator David Durenberger, a German Catholic from Minnesota, fudged on a condominium deal for between $3,000 to $5,000. The lawyers tied him up endlessly in court and forced to accept a probation sentence of one year.

IRISH PSYCHOLOGY/IRISH PSYCHIATRY: THE TIRESIAS COMPLEX

Ivan Boesky, a New York stockbroker, stole or swindled 300,000 to 700,000 million or more from the working poor in this country and got off with practically no sentence at all. Boesky is one of the biggest scourges to the poor this country has ever had. Is this justice? How would you resolve this paradox?

David,

Mary Ann O'Dougherty and her band, Mandala: Five CDs

My sister, Mary O'Dougherty, has another musician, Tim Sparks, in her band who is having a CD release at the Cedar Culture Center March 29, 1996, at 8:00 p.m. It will be his third release. His previous two releases are Guitar Bazaar and Nutcracker Suite also containing Balkan Dream. Terence Hughes the pianist in her band has a CD entitled East of the Sun.

Patrick A. O'Dougherty, Ph.D.

David,   2/19/96
Deep Blue Records
If you have an interest, my sister Mary Ann O'Dougherty's digital release, Water Color of Brazil, is Friday and Saturday night at Dakota Bar in Bandana Square at 9:00 p.m. Her band is Mandala.
Patrick's Life in Psychology: The Invention of the Fields of
Psychology/Psychiatry Noir
Beyond Capitalism
Have you read Frederic Jameson's, Postmoderism or The Cultural Logic of Late Capitalism? He is a Marxist at Duke University. I think America should have a scientifically, mathematically, theologically based economic system.
DFL: "The War Party"
The Democratic Party in America is the "war party." For example, O.J Simpson "Balkanized" America. The reason is promotions in the military or in the government go slowly during peacetime. The Republicans want to stay home and make money, open trade routes or arrange diplomatic agreements.
Liberalism and Anti-Catholicism
The big enemies of the Catholics in the "last bastion of anti-Catholicism the American school system" are the liberals. There really isn't much of a military option without the Catholics or the Church. Americans aren't willing to die for the idea they are "manure."
Genetic Research
One of the biggest coming areas of research in this country is genetic engineering. It will change genetic warfare. Is it a byproduct of the Civil Rights Movement? Women are genetically superior to men. If a child dies in childbirth,

more often than not it is the male. People have known this since the ancient world. This is a big area of research in the women's movement. Many people claim homosexuality is genetic. People learn most of it; and, I didn't learn it.

St. Paul/Minneapolis: Sodom and Gomorrah

St. Paul and Minneapolis are the Cities of the Plains. The major areas of the homosexual movements in America are San Francisco, Minneapolis/St. Paul, New York and Washington,D.C. The homosexuals have the DFL in Minneapolis tied up. The social structure of the Twin Cities has a pyramid character.
Patrick A. O'Dougherty, Ph.D.

David Noble,   2/14/96

The Irish: A Cyclical Counterpoint and Tonic Key to the Austrians
Patrick O'Dougherty on The Irish and the Austrians

My mother, Patricia Coyne, got me a scholarship to the German Kulturhaus in St. Paul to listen to William Wright lecture on Austria. So doesn't feel comfortable with the Windsors and England. I think Austria is the way for Americans to develop in parallel. Austria is a strong Catholic country. Minnesota is the main state in America for German and Austrian identity. Also, Karl Popper and Jewish identity came out of Austria. The Austrians are great at music and so are the black people. Vienna is a city that defines cosmopolitanism. In contrast to Austria, in the Middle East, Africa and the Orient is there are great extremes. The only conclusion most of the people of Europe, for example the Irish and the Austrians, could come to about these areas is that religous synthesis should originate there. Let the Christians and the Jews sanctify through space and time Patrick O'Dougherty's green, grey and the purple revolutions. For example, his personalist intuitionist philosophy is a counterpoint revolution to Wittgenstein's philosophy of language.

Patrick A. O'Dougherty, Ph.David.

David Noble,   2/21/96

Sartell, Minnesota, Courage

I have some small town cousins, from Sartell, Minnesota, named Frank and Marilyn Rajkowski. Frank is my cousin. He is a lawyer with five children. Their daughter, Annie, took a class trip to New York to see some stage plays.   Annie decided to go to Rutgers University and major in acting. Don't you think that takes a lot of "Mother Courage" for a small town Polish, Irish, Norwegian, Catholic girl who has rarely been out on a date? How do you rate this effort as a professor?

The Democrats invited me to leave their party,

IRISH PSYCHOLOGY/IRISH PSYCHIATRY: THE TIRESIAS COMPLEX

Patrick A. O'Dougherty, Ph.D.

David,   2/28/96
Issues in Hydrogen Conversion
The big issue in economics is hydrogen "conversion." The issues are fuel from water, water from fuel, desalination, the Hindenburg Blimp, and sea water. Hydrogen composes ninety percent of the universe. It is a pure recyclable fuel. Most of the water in the world is sea water. Do you have any ideas on how to approach from the topic from the perspective of history? Petrochemical importation is the big problem in Minnesota's economy. It is big money. Russia is a landlocked country. America, England, Ireland and Israel are not. This is a major defense issue. Electrolysis releases it from water. The body is ninety percent water. Two gallons of hydrogen fuel in an automobile will drive it 300 miles.

Patrick A. O'Dougherty, Ph.D.

Maureen,   3/9/96
"Up From Slavery"--The Defeat of Slavery
Like St. Patrick, I was a slave in the South in 1974. I got eight months on Bellwood Chain Gang Atlanta, Georgia, the City of Martin Luther King, for a very minor offense.
Patrick A. O'Dougherty, Ph.D.

David,   3/4/96

Males or Females with Very Fragile Egos
How do you deal with males or females with very fragile egos? Many professors do not have this type of personality. However, some do. If you had a son or granddaughter who lost a softball game, would it be too much for you? Does Gail have you walking on eggs? Life is a pecking order, right? I think the idea is too deal with this type of person very infrequently. If someone doesn't agree with you, does it scare you? Can't you discuss differences rationally? I think this type of personality gets way in the way of productivity. I think my feet are on the ground.

Patrick A. O'Dougherty, Ph.D.

David Noble: 3/25/96

Divide and Conquer
I am not trying to be offensive about this, however, I never considered converting to Protestantism or Judaism. Catholics think there is unity and coherence

196

in God. There may be twenty-five hundred to twenty-five thousand Protestant sects. These sects are always disagreeing, killing each other, and emphasize irrationality, individualism and fundamentalism. Is this God? Cheryl Locketz, a Reformed Jew, told me modern Judaism has "no conception of an afterlife." How are you responsible for your actions then? Martin Luther and Henry VIII brought all this vicious nationalism into Western Civilization which has been in almost a constant state of war since the Protestant Reformation. Few Europeans are real wild about English Protestantism. The English will probably champion most of the anti-Catholic causes in the European Union. They will be a small minority in that union. Brother Gregory Conant said, "Not one Protestant sect many any formal concessions to the Catholics as a group."
Dialogue not conversion,

Patrick A. O'Dougherty, Ph.D.

David Noble: 6/12/96

The Southern Mind: A "Siege Mentality"
Newman Center: A Case Study

Many of the Northerners are naive about how Southerners think. They have a so-called "siege mentality" about everything in the North. This is especially true in the Deep South. I lived there for about a year and a half. For example, they placed a group of Southern uplanders on the staff at Newman Center. They have subtly been playing the race card at me for five years. I get no rights, no say, no dissent, negativism, and nothing. Am I supposed to go for it? I don't go for it. Would you? I am having a little legal rhubarb with the staff over at Newman Center about this. I want to develop the hydrogen conversion project and the alternatives to petrochemicals project to give some balance to the Arab oil monopoly. For example, many Arab people do not benefit from the oil trade. I'm from a 100 year family in the State of Minnesota and it saddens me to see Minnesotans taking on a Southern siege mentality.

God Bless David Noble,

Patrick A. O'Dougherty, Ph.D.

David Noble: 10/4/96
Paradoxes of English Protestantism

In England, they have signs in different pubs which state: "Dogs, Irish and Negroes do not enter." Is this Christian? Oliver Cromwell went into Ireland and martyred 300,000 people. Is this part of being a "race of servants"? In England, they use the Irish for social control purposes of the Negroes, Jews, Muslims, Indians, and women. What are some Biblical passages that would back these types of policies?

In England, they traditionally also had several clubs exclusively for men. What is going on with between the men in those clubs? Henry VIII founded the Church of England. Don't you think that this is a calling card to the Irish, the French, the Austrians, and to the Italians? Why don't they found a Church of Minnesota here. Few of the Europeans convert to Protestantism because they can't understand how God would have many warring sects in his mind. Catholics bring in about one third to one half of the trade to America. Is this an example of what JFK meant when he said "life is unfair?"

Small Town Parents

My parents are from very small towns. Their son wanted to build up the nonwhite people and create a classic. The father was a military man from Maple Lake, Minnesota. He was intelligent. However, his blind side was that he was naive and trusting of people. He thought he would give his son an education that money couldn't buy. He gave him all kinds experiences in the inner cities.

Patrick A. O'Dougherty, Ph.D.

Paul Murphy,   6/19/96
Juneteenth: Through a Glass Darkly

I have built an internationally recognized library in sustainable agriculture in the Dorothy Day room at Newman Center. It has thousands of documents. I have founded a new school of mathematics and physics--the personalist intuitionist school. I have published seven books, and I have an archive to build into a psychiatric archive. One of my intellectual goals in life was to create a black classic so I choose a famous day in Afro-American Freedom History, Juneteenth to write you.

The Deal on the Intellectual Threats

When I was growing up in the 1950s and 1960s, there was a lot of racism in many of the Churches. I'm not for racism in Church. I find that to be offensive. I am not a nut about sex. However, I find pornography in churches, like that pornography forum you were a member of in Macalaster's chapel to be very offensive. For example, the staff at Newman Center has tried blasphemy, swearing and quasi Marxist liturgies. My position is that the constitution does not protect the right to introject the profane into Catholic liturgies. Also, Archibald Cox came to the

IRISH PSYCHOLOGY/IRISH PSYCHIATRY: THE TIRESIAS COMPLEX

Presbyterian Church in downtown Minneapolis and ended his sermon there with the high sign. Do you think this is right for little kids or Christ? A big area in mathematics is boundary value problems. This should hold for religion also. I've really never been violent except in self-defense and "the past predicts the future."

Black Irish

Black Irish is a term of opprobrium for somebody who betrays the faith and the Irish people. Is this a psychological problem for you? I'm not an atheist/agnostic who has spent his life championing free speech and the pornography movement. None of my friends are for you. There are not a lot of Irish at the University of Minnesota, and I do not want to make this my mark. Did Kennedy, McGovern, McCarthy, Joyce or Donnelly do this? The term, "Black Irish," is also kind of a positive term for the dark featured Irish as opposed to the Scandinavian light featured Irish. Also, I feel it is a form of tyranny to tie oneself to one constitution. Most of the Europeans did not do this. I look to the Middle East, Africa and Ireland for a new contribution to soul. I personally think you would get along a lot better with everybody including yourself if you did this. Will you save your soul? I hope that I get credit for my positive contributions in life.

Evil

For some reason life typifies by social conventions, suffering, particularities, and evil. Lucifer was the most intelligent angel. His sin was "non serviam." He would not serve God. Christ didn't come into the world to condemn it, but to save it. However, half the numbers on the number line are negative, so maybe the saved just equal the ones not saved. I hope Christ saves Paul Murphy "through a glass darkly" which is a theme from St. Paul.

In the spirit of forgiveness, through a glass darkly

Patrick A. O'Dougherty, Ph.D., A Juneteenth black sheep
P.S. David Noble is a professor friend of Paul Murphy.

David Noble,   9/27/96
The Baby Boomers

Historians don't give any American president in my lifetime good grades except Truman. How do you rate the Baby Boomers? They did the Civil Rights Act, Vietnam, the sexual revolution, drugs, the women's movement, and the Americans with Disabilities Act. They are fairly "ordinary" historical contributions. I think peace studies is going to develop on many campuses.

Legal Case

I had a very minor civil rights case with the Italian priest, Father Steve Bossi, at Newman Center. He moved on. The hold on the case is up November 13th.

IRISH PSYCHOLOGY/IRISH PSYCHIATRY: THE TIRESIAS COMPLEX

Business Deals

I made some good money for our family--13 years in business. However, the Church teaches <u>money is a false God</u>. <u>I am going to put my books on Internet</u>. We came out on all of our business deals except the law office which we will probably come out on.

Movements in History

I like teaching, but I also like the movements in history. I was a charter member of Student's for a Democratic Society on campus. I did the Irish white Negro activist for writing definition in graduate school. I helped found a Benedictine Oblate group while I lived in St. Paul. I was a member of the Legion of Mary for eleven years. I'm a psyche activist, and I'm a member of the Green Party. I'm considering doing a Hibernian critique or ethnic critique of the Green Party and an ethnic critique of capitalism.

F. Scott Fitzgerald

Mark McGee and I went to the F. Scott Fitzgerald 100th birthday party Wednesday at the Landmark Center. Garrison Keillor signed my program. The city issued a stamp in memory of Fitzgerald. There were speeches by Garrison Keillor and Mayor Norm Coleman.

Mentoring

I am an alumni and could become a mentor at the university.

Patrick A. O'Dougherty, Ph.D.

David Noble, Anglican

9/30/96

A New Covenant in Minnesota American History

I went to the Evensong ritual celebrating the Anglican-Roman Catholic Covenant at the Cathedral Church of St. Mark on September 29, 1996. The ritual brought me into consonance with your covenant writings in American history. We have a covenant relationship. This act in Minnesota history will renew the face of the earth. I especially thought of our covenant with the Native Americans at this signing. Here is a copy of the Anglican and Roman Catholic Covenant and the Evensong liturgy.
Please retain for your files.

Patrick A. O'Dougherty, Green Party theoretician

David Noble,

IRISH PSYCHOLOGY/IRISH PSYCHIATRY: THE TIRESIAS COMPLEX

10/4/96

Conant, Freeman, O'Dougherty

My friend from Massachusetts, Brother Gregory Conant, OSB, was in town. I took him to have my therapist analyzed him. He had never been to a therapist before. He talked about how he had reconciliation with his father before his father died. Brother Gregory has pressured speech, religiosity, and was a prisoner of conscience. I brought him up to St. John's Liturgical Press so he could present his ideas for a book in history. He calls it "So You Don't Like History." They liked his topic real well at the Liturgical Press. Father Hilary Freeman, Ph.D., sent me some material on Christian freedom which I requested. I am going to preface my book on my letters with this theme. I have helped Father Hilary get published. I am taking some time this quarter to address the personal validation issues of an almost fifty year old single man. My book on the Green Revolution is done. The title is St. Patrick, The Green Revolution, and the Hydrogen Conversion Project.

God Bless,

Patrick A. O'Dougherty, Ph.D.

LETTERS TO TERRY GIPS
The Hellenist America Institute

Terry Gips,   12/30/93

In recognition of your contribution to the field of Agricultural Economics, specifically your books on Breaking the Pesticide Habit and on The Humane Consumer and Producer Guide and for your newsletter, Manna, the Graduate School of the Hellenist America Institute--University is going to award you a doctorate of letters degree. You have made an outstanding contribution to the field of sustainable agriculture and your contribution should not go unrecognized. Today is not only the occasion of the ratifying of the Vatican-Israeli recognition of the State of Israel, but it is also a benchmark day in the area of Hellenist-Jewish relations. One can only hope that a reinvention of the Hellenist-Jewish synthesis will result.

A degree certificate will be forthcoming.
Sincerely,

Dr. Patrick Aquinas O'Dougherty
President of the Hellenist America Institute--University

IRISH PSYCHOLOGY/IRISH PSYCHIATRY: THE TIRESIAS COMPLEX

Patrick Aquinas O'Dougherty, Ph.D.
Riverside Plaza M4310
1615 South Fourth Street
Minneapolis, Minnesota 55454

John S. Pillsbury Jr.
1280 Bracketts Point Rd.
Orono, MN

Sir: 8/14/95
The Lack of Idealism on Campus
My father, James Aquinas O'Dougherty, was an attorney for you at
Northwestern National Life Insurance Company during the 1960's. He always
thought quite highly of you. I am a historians/activist at Newman Center on campus.
John Henry Cardinal Newman was one of my favorite Christians. I have an interest
in getting a priest/scientist, like John Newman, on the faculty of the University of
Minnesota to build up the idealism of the young people. I have an involvement with
the International Alliance for Sustainable Agriculture which is in the Dorothy Day
Center in the basement of the Newman Center. A Yale man, <u>Terry Gips</u>, co-founded
the alliance. The University of Minnesota Episcopal Church also used to be at
Newman Center. They moved across the street. I have built a research libraries in
sustainable agriculture in the world at Newman Center. I have an interest in having
this group apply for a Nobel Prize. I want to establish a new Noble Prize category
for women's contributions. They are fifty-one percent of the world's population, and
they deserve it. What do you think of these projects? I got my first summer job at
NWNL. You can't have a better starting point than that can you? I would like to
refound the University of Minnesota.
God Bless,

Patrick A. O'Dougherty, Ph.D.

4/25/96
Terry Gips,
Blackout,

Patrick A. O'Dougherty, Ph.D.
<u>**"The Irish Dogs, Irish Scum, Irish Micks"**</u> These phrases were common racial
slurs of the Irish. The English still use them.
The Irish White Negro Issue At Newman
According to Dick Hope, back in the 1950s in New England, there were signs
saying "Dogs, Irish and Negroes Do Not Enter." They <u>still</u> have signs like this in

England. Do you think Martin Luther King liked this? The 1960s were the end of the "Dark Ages" in America. The first Catholic president and the Civil Rights Bill ended America's "Dark Ages." Also, the Green Party leader, Petra Kelly, died in a murder suicide in Germany.

cc. Terry Dosh and David Noble

Terry Gips: 6/19/96 **Juneteenth**
        The deal on Newman Center in the legal hearing was that if I agreed to drop the lawsuit I get to keep doing the movement and library there. The legal hold is up in November. I did a lot with Lenore Scallen and with her library also. I have an interest in starting a Minnesota Psychiatric Archive now. I have one copy of the books in the IASA movement. I could move on. However, I don't have a problem with Terry Gips. Father Steve and Father Rich gave me backhand invitations to involve myself with them.

Terry Dosh
        The deal on Terry Dosh is that he never did anything with his doctorate. The intellectual problem that I have with him is that he doesn't have either a major or a minor work and his newsletter is mainly clippings from other newspapers. Also, he wants to undermine the status of the celibate priests, and I'm not for it. I feel he has the queen bee and drone syndrome. His wife supports him. I would not feel comfortable with that.

My Ecology Book
        What was Todd's last name who built the IASA file list? Who worked with him on it. I want to credit them academically in my book. I will get the references from Leo Cashman. I am going to have a partial printout of the resource center in this book because it will enhance the book and my contributions to the resource center endure.

Terry Gips
        I wasn't raised to be prejudiced, and I don't want anything to do with racism or discrimination. I have dealt with all kinds of people in my life. Let's deal with Terry Gips. Today is a very famous day in Afro-American History. Did you ever have a male punch your ticket? It happened on Juneteenth, June 19th., 1996.
        God Bless,

        Patrick A. O'Dougherty, Ph.D.

Terry Gips,   8/6/96

A Hydrogen Powered City in Minnesota

IRISH PSYCHOLOGY/IRISH PSYCHIATRY: THE TIRESIAS COMPLEX

Minnesota must be one of the first states in the country to have a hydrogen powered city. I guess there is a movie out called "Chain Reaction" about hydrogen. I'm going to see it. Hydrogen composes ninety percent of the elements in the universe. It is a life and a death issue for Ireland, Israel and China.

The Chemical Mapping of the Brain

I've been given an invitation at HCMC to do a book on the chemical mapping of the brain. It is the key to solving many medical problems, for example alcoholism, depression and sex problems. Maybe a chemical map of your brain would solve your problems. I am also planning to do a book on alternatives to petrochemicals in medicine. There are many.

Newman Center

I heard that both priests are transferring at Newman Center. I liked Father Richard Colgan. However, the point I raised with him is that he can pitch but not hit. For example, he gives good sermons. He does good work with the gays. He does his homework. However, he hasn't published. This is the reason Newman Center isn't that well accepted on campus. "Publish or Perish." Has he produced a vocation? Is he good at administration? I hear that he wants to try the big leagues, Boston. Why not try the "B" squad or a farm team if you aren't going to go for it? Boston is the hardest archdiocese to get a promotion in. Father Steve Bossi has many strengths. However, he was always subtly negative with me. I hope that this legal situation and stay of evaluation resolve the problem. I am going to emphasize his strengths and hope this resolves the problem. He is cool and weak on psychological mindedness. I am not going to rub this in. I want to continue building the library after this legal deal is up after November 13, 1996. Steve and I are not to contact each other during this period. This is often done in legal hearings. After this, I could move on and focus on Harvard and St. Benedict's College. I want the summer to be a period of growth for me. I ended up having some Viktor Frankl experiences out of addressing an Irish white Negro experience in Atlanta during 1974. Did you ever have any Viktor Frankl experiences?

My Ecology Book

I want the names of the student interns who helped prepare the list of organizations and files for the IASA resource center so I can footnote their names. I am going to include the data base as part of my book on St. Patrick, The Green Revolution, and The Hydrogen Conversion Project. I am going to footnote this the way that they tell me to at the literature section of the public library. They are experts on style manuals. I think you will like my book.

Terry Gips

Where does Terry want to go on his journey or odyssey in life "from where the sun now stands?" This is a famous peace quote from American Indian lore. Do you recognize it? Maybe you want to try something completely different. You are my

Manna in the desert. Also, I like the Radical Republican tradition of Thaddeus Stevens and Charles Sumner. I'm not a pacifist and vegetarian after what happened to the Irish.

Talk to you soon,

Patrick A. O'Dougherty, Ph.D.

LETTERS TO HIGH SCHOOL FRIENDS

Patrick O'Dougherty, Ph.D.

10/16/94

Sara Bey Dreyfus Samson
1412 Lee Road
Northbrook, IL 60062

Life: Proving An Incompleteness Theorem

Sara,

I 've given you an intellectual and artistic defeat. How about a spiritual defeat? Goethe thought that God is "the eternal feminine." Why don't you prove that theory? Have you ever read Jo Freeman's, "The Building of the Gilded Cage?" Have you read Kate Chopin's, <u>The Awakening</u>?   Why don't you write a reply to Schiller's, <u>The Aesthetic Letters</u>? In response, you could define or create a non-repressive order in America--a revolution starting with yourself. I've invented a Marxian theory of marriage. It can be a dictatorship over the proletariat--the women and children.

I have an interest in trying to refound world culture. I want to do something Irish--write a classic.

Women are not like bunnies or rats. They are qualitatively different. Hugh Hefner has a second rate mind.

I hope you save your soul. I hope this has been helpful.

A Double Bind,

Pat,

12/7/94

205

IRISH PSYCHOLOGY/IRISH PSYCHIATRY: THE TIRESIAS COMPLEX

Patrick A. O'Dougherty, Ph.D.

Frederick J. Schroeder
6224 Garnett Drive
Chevy Chase, MD 20815

Frederick:

Sorry you cannot come to our 30th. Washburn class reunion. Specifically, I bank at Riverside Bank on the West Bank in Minneapolis. They give the customers at this bank 2.25% on a $1,000 minimum savings account. I have a friend named Thomas Linstroth who is a friend from St. Thomas Apostle Grade School. Like ours, his family is worth quite a lot of money. He tells me that he gets 7.5% on his five year Treasury Notes. I think the customers at Riverside Bank have Alan Greenspan to thank for such low returns on their accounts. **I think the president should fire Alan Greenspan for this.** I wrote Hubert Humphrey about Henry Kissinger, and I put a good word in for him. I have taken many economic courses, and I think that they are doing such a bad job with economics in this country that what they should do at the Federal Reserve is reissue Confederate currency.

In the interest of bank competition,

Patrick A. O'Dougherty, Ph.D.

12/7/94

Patrick A. O'Dougherty, Ph.D.

Sara Bey Dreyfus Samson
1412 Lee Road
Northbrook, IL 60062

"The Light Touch"

Sara,

I have done a lot of teaching and writing, and you have to be somewhat tough on the students. You are not a historian until you've lost a battle, right? I hope that I wasn't too tough on you. Anyway, there are many good ideas about high school education in America, however, do you think there is much substance to secular/humanism or a basically atheistic/agnostic school system like that at Washburn? It is a "manure philosophy," right? How can you go against the experience of the death of God? What do you think about the ideas of "no concessions" and "see you on the other side?" I have had the same goal since I was at Washburn--to produce a black classic. Do you have a spiritual dimension?

Verstehen,

Pat

Patrick A. O'Dougherty, Ph.D.

Sara Bey Dreyfus Samson
1412 Lee Road
Northbrook, IL 60062

Sara,   10/11/94
                               Artistic Defeat
          What is wrong with giving a woman an artistic defeat? I was over in Paris and I accepted Simone de Beauvoir's defeat, a woman's defeat. My mother and sisters and I have published a lot. I am curious why you shifted into autopilot? There is a German nun up at St. Ben's who writes me. She is formidable. Have you ever read Thomas Mann's, <u>Dr. Faustus</u>? Adrian sells his soul for musical genius and then doesn't get it. This happened in America. Where is the Bach, or Brahms? Have you ever thought about redefining German mythology, the "myth of German superiority," for example. There are several good myths about the Germans that one could develop. Myths are a major part of a peoples' identity. I'm single and celibate. My institute is named after a woman, Helen. I didn't click at Benilde, but, everybody at Washburn liked me. I'm sending you an article that I got published in the <u>Minnesota Daily</u>. I slowed this Young Turk, Chris Fellerhoff, down.

With empathy and respect, "big prize, big price,"

Pat

Patrick A. O'Dougherty, Ph.D.

                                          1/5/95
Sara,   **For Starters**

          First, I want to try to convert the world to hydrogen use. Remember the Hindenburg. This accident is the reason given to the American public for not using hydrogen in automobiles. It is the dominant element in nature, the most powerful and efficient fuel, and the most clean. Most countries do not have fossil fuels and it is going to be a major mistake to use them up. They can't be easily replaced. Second, I think there should be a new Nobel Prize category started for women's contributions. They make up fifty-one percent of the world's population. Third, Americans deny Christians a tremendous royalist heritage, and I want to recapture and redefine it in the New World. Fourth, there should be a Holy City in the New World--St. Paul. Fifth, there should be a new royal family in England. Princess Diana's aptitude scores were so low that they weren't going to let her graduate from high school. This shows a below average I.Q. My mother thinks that the Windsor children are all very average.

I suppose that you have many strengths. Do you still work for World Book Year Book? Why don't you fax your material to the Nobel Prize Committee to see what they think? The Irish have lost millions. It is not a game with an Irish Catholic. Women have lost billions. Are you one?

To mystery,

Patrick A. O'Dougherty, Ph.D.

Frederick J. Schroeder
6224 Garnett Drive
Chevy Chase, MD 20815.   5/5/95

Happy Cinco de Mayo. First, some Catholics want to have a "culture war." I think that they are going to win it. Second, have you ever read Thomas Mann's, Buddenbrooks? His father was German, and his mother was Latin. Buddenbrooks is about the decline and fall of a German family, like the Windsor family, over four generations. I think that it could be the Washburn story. Third, intelligence can be a downfall. For example, Faust sells his soul to Mephistopheles for knowledge and power. Did you do that? Fourth, when I went to Benilde, the assistant principal, Brother Patrick, was going to flunk me out for a minor religious infraction. I wrote an outstanding essay in response to this.   They encouraged me to stay on. However, I decided to switch to Washburn High School. The principal of Benilde, Brother J. Ferdinand, told me that it was a mistake to transfer to Washburn High School. He thought I would do better than most of the students there.   Brother Ferdinand couldn't think of a single name they had at Washburn. He didn't have an axe to grind. Washburn didn't impress him. He was a leading Catholic educator in the state. Do you think that he was right? Fifth, Beethoven thought "none of his teachers taught him anything."(Reference: John O'Dougherty) Was that your experience? Sixth, at St. John's University, they had us read F. Hayek's, The Road to Serfdom? The theme is that big government will lead to totalitarianism rather than a bountiful harvest. Do you think that the Federal Reserve Board will force a totalitarian takeover? Are you a serf in the bureaucracy out in Washington? Hayek's theme has personal implications.

Patrick A. O'Dougherty, Ph.D.

Frederick J. Schroeder
6224 Garnett Drive
Chevy Chase, MD 20815

6/28/95

Fred:

      A major theme in Irish-Catholic literature is the theme of the queen bee and the drones. This theme has become a central problem in American history. Is it true for the field of economics? Is this the problem at the Federal Reserve Board? I'm sending you some material on a Chicana writer. Enjoy it!

Patrick A. O'Dougherty, Ph.D.

Patrick A. O'Dougherty, Ph.D.

Frederick J. Schroeder,   1/12/95
6224 Garnett Drive
Chevy Chase, MD 20815

      Have you read William Greider's, Secrets of the Temple: How the Federal Reserve Runs the Country? Is the Federal Reserve a Temple? Is it a shadow banking system? Is it Fred under Karl? Is the Federal Reserve a conspiracy? Is this book truth or trash? Did you ever think about faxing your research over to the Nobel Prize committee that looks that economic contributions to have them rate your research? I think there should be a new Nobel Prize category for women's contributions. I sent a letter to this effect over to the Nobel Foundation in Stockholm, Sweden. I'm a royalist/radical economist. I liked the economics of Frederick the Great of Prussia. I like radical/personal economics, like the economics of women.

IRISH PSYCHOLOGY/IRISH PSYCHIATRY: THE TIRESIAS COMPLEX

You're not an economist until you lost some money. How much money have you <u>lost</u> or made? Have you ever thought about making a name for yourself by prosecuting a Washington figure out there?

Break the conspiracy,

Patrick A. O'Dougherty, Ph.D.

Sara Bey: 6/16/96
                                    Bloomsday/1996
               A Spiritual Defeat: The Spirit of Washburn/1964
I will apologize for being a little tough on you with a spiritual defeat. However, what is the name, Sara, a symbol of? Isn't it the spiritual creation of the Jewish people? Secular education, like that found at Washburn High School, has a lot of strengths, especially academic variety. Its pitfall is that it lacks balance. Does one give a very materialistic civilization, like America an "A"? Materialism is often a denial of freedom. Like Sara, St. Patrick was the saint who rose up from slavery to a spiritual and free definition of the Irish people, the people who "God made mad." I hope that I transcend all forms "of human bondage" which is a theme from Spinoza and Maugham. The reason that I'm writing you is that I wonder how would pull the class of Washburn 1964 out of the bondage ordinary, capture its spirit and idealize it in time? For example, Bloomsday is a day that got pulled out of the context of the ordinary. This is also true for the Bloomsbury Community in London.

An Irish Catholic in German Washburn High School/1964,

Patrick A. O'Dougherty, Ph.D.
See current listings of my publications and research library.

                                    5/31/96

Frederick J. Schroeder
6224 Garnett Drive
Chevy Chase, MD 20815

Frederick:

                Banking and the Impossibility Theorem

IRISH PSYCHOLOGY/IRISH PSYCHIATRY: THE TIRESIAS COMPLEX

Kenneth Joseph Arrow, economist, has a provocative theorem on economics and government. He proves mathematically that a perfect economic government is not possible. Is this true for the Federal Reserve Board? Is this true for banking competition? Is this true for currency demand? Also, do you think that Bill Clinton is America's first Marxist president?

An Agriculture Economics Library

I have built a major International Alliance for Sustainable Agriculture Library at the Newman Center on Campus at the University of Minnesota. I am sending you a copy of its data base.

Washburn High School: Irish Culture

I did have a good experience at Washburn High School compared to Benilde. However, I am a Catholic dissident and want to through some cold water on secular education. Washburn is "this side of paradise" to quote F. Scott Fitzgerald. I have an Irish friend named Mark McGee who grew up on 45th and Garfield who I went to Benilde with. I am sending you his Curriculum Vita for comparison.

Patrick A. O'Dougherty, Ph.D., An Irish Catholic in German Washburn High School/1964

Sara Bey Dreyfus Samson
1412 Lee Road
Northbrook, IL 60062

Sara Bey,   10/1/96

I would give you any more criticism or negativism. I apologize for to you for it. Instead, I am going to give you a challenge to excellence. I taught writing at the university. So, I am sending you a woman's writers list. I pray that you come up to their level. I've become a theoretician and activist with the Green Party. I am sending you some material on it. Winona LaDuke, a Minnesota American Indian women, is their vice presidential candidate. Also, I gave in on everything over women.

IRISH PSYCHOLOGY/IRISH PSYCHIATRY: THE TIRESIAS COMPLEX

God Bless and vote the Green Party,

Patrick A. O'Dougherty, Ph.D.

Sara Bey,   12/7/96
Apologies, Positives for Sara, and Moving On
Apologies
   I apologize for being pretty tough on you there. However, a Catholic girl, or a Afro-American girl or an Oriental girl or a Jewish girl could have become the Presidential Scholar and gone on to Yale. I hope with breaks like that you come up with a lot. I cut most of the criticisms and the negatives from my files. My mother and sisters are all somewhat famous and are for being very tough on me professionally. Maybe I was just returning the fire. The reason I raised the queen bee and the drone issue is because I'm not pro choice. I'm pro life all species. I apologize for raising this issue. I am sending you some material on Margaret Ward's, The Missing Sex and on her work, In Their Own Voice. I hope you rise to her challenge and refute the '"collective amnesia"' of males.

Positives for Sara
   I remember Sara as being intelligent, good looking and kind of classy. She was good at writing and chess. She ended up being Ivy League educated. I had a good sense of humor in high school and the Biblical quote that comes to mind is Genesis 21:6 And Sarah said, "God has made laughter for me; every one who hears will laugh over me." I hope my response invites laughter. Also, one of the most creative acts one can do in life is to become a mother. Like Sarah, you became a mother when you were older in life. Will you come up to the level of the Biblical Sarah as a mother.
Moving On
   I am sending you a copy of one of my books. It came out of this correspondence. I hope you find it to be a spiritual treat. It contains the data base of the library that I built. It came out of the Washburn experience. I don't know if you believe in God. However, God is "self-gift." This book is my self-gift to you. I'm moving on from this relationship.
God Bless,

Patrick A. O'Dougherty, Ph.D., Irish white Negro, revolutionist

Frederick J. Schroeder
6224 Garnett Drive
Chevy Chase, MD 20815

IRISH PSYCHOLOGY/IRISH PSYCHIATRY: THE TIRESIAS COMPLEX

Fred, 12/18/96
The Green Revolution
I wrote the enclosed book, <u>St. Patrick, The Green Revolution, and the Hydrogen Conversion Project</u>. The Green Revolution and agricultural economics relate closely to your field of economics. Why don't you give it a critical reading? It came out of the Washburn High School experience. Also, here is a copy of the State Party Platform of the Green Party of Minnesota which is a fulcrum of the Green Revolution.

Washburn
I basically liked the students and teachers at Washburn High School. The strength of the secular school system is diversity. Its downside is that it breeds religious indifference. Did this happen with you? I hope the religious challenge I gave you bears fruit. An intellectual should be provocative and somewhat offensive. I raised the queen bee and the drone syndrome with you because I think it relates to the abortion issue. I am pro-life all species. I enjoyed meeting you at the twenty-fifth Washburn Reunion. I am going to move on from this relationship. It did, however, bore fruit, for example, the book. Did you ever think about publishing something in economics? The dictum at Yale is publish or perish.

God Bless,

Patrick A. O'Dougherty, Ph.D., Irish white Negro, Green Revolutionist, out of Mpls. and the Harvard Benedictines.

Vance Opperman: 9/14/95
Women Writers/Artists on the Left
I wrote you because I wanted to find out what happened to Howie Kaibel and Lee Warren Smith. Also, I have an interest in writing a book about women writers/artists on the left. This is something you know about. For example, I met **Madeline Hammond** at a party thrown by Milt Schoen during your campaign. I had an interest in having some down and out in Minneapolis and St. Paul experiences, and so was she. She was a small town girl from Fergus Falls, Minnesota. She was the first child from her family ever to go to college. Her family got her cello lessons, and she majored in cello at the University of Minnesota. However, she recorded an original composer's piece, Dr. Good's, at the Walker Art Center after her graduation recital. I went to the recording. I thought the contribution was pretty impressive for a small town girl who wanted to do big times with me. I called her up last summer, and she is still a single mother. She has three daughters. She has an M.A. in cello, works in a violin shop and plays in a symphony. Another girl I went out with while I was in Students for a Democratic Society and the Young Socialist Alliance was **Theresa Lippert**, a lawyer, who works for you at West Publishing. She was a Regina

IRISH PSYCHOLOGY/IRISH PSYCHIATRY: THE TIRESIAS COMPLEX

High School girlfriend of my sister's, Peg. I had heard my sister had a girlfriend who wanted to be a writer. I answered the phone over at my parent's home on Gladstone one night, and it was Theresa. She talked to me for forty-five minutes. She wrote me a thirty page letter the next day. I asked her if she would consider going out with a college student. She did. My parents were friends of the Lippert's though Gene McCarthy's campaign. Robert Lippert taught English at the College of St. Thomas. My parents would have never approved of the relationship. It was too close for comfort. They didn't approve of my career choice either. I wanted to write a "black classic." Later, I called up Theresa and asked her what she wanted to do on a date. She wanted to see the movie, "A Man and a Woman" at the Grandview. I went over to their home on Cretin avenue and met her dad, and we went out on the date. She told me her dad liked To the Finland Station by Edmund Wilson, one of Mary McCarthy's husbands. The relationship blossomed for a while. For our second date, there had been a terrorist bombing at the Federal Building in downtown Minneapolis. Remember it? She had me take her down to the Federal Building and make her a love necklace out of the debris. We got **beyond the Finland Station**. We broke up after a while, but you wouldn't forget an experience like that would you? I had an Irish psychologist for a counselor at the University of Minnesota named Dr. Shirley Corrigan. When the draft issue for me came up during the war in Vietnam, I didn't feel comfortable with it. Robert Kennedy's assassination had just occurred, and I didn't feel comfortable with the officers from "another country--the American South." I was not a pacifist, but they didn't want a young Trotskyite in the military. I told her about this. She suggested I have the draft board call her in her office. They did and she told them no dice, and that I was an intellectual and not the type of person they wanted in the military. She had been a military nurse during WWII. I had to make a malingering/psychology case at the draft board and that ended it. Shirley was part Jewish too. The draft board never questioned her judgement. For Pat it was "a separate peace." Thus on a personal level an Irish Catholic and an Irish psychologist won the war. When I was in graduate school, my uncle, Terence, told me that if you wanted to be a writer, don't do textbooks or experts. Writing was not his calling. He thought what you should do is to take a small amount of money, head out to a destination, and see what happens. My dad liked Thomas Wolfe's book, You Can't Go Home Again. Wolf is a famous Irish Southern writer. After a psychological underworld experience, I started hitchhiking, ("You Can't Go Home Again") down to Atlanta with about thirty-seven dollars on me. I wanted to reconstruct ancient civilization in the South. All hell happened to me. I had written **Caroline Kennedy** during my senior year at the University of Minnesota, not to have a romance, but because I thought she was at a very vulnerable age for a young girl growing up with a slain father. Anyway I ended up getting busted in Atlanta for a **very** minor infraction. My dad decided to stay out of it. He had been a drill master in Officers' Candidate School in Maryland and a former attorney for John Pillsbury. He wanted to get me toughened up. So he let a Southern judge sentence me to

Bellwood chain gang for eight months. In effect, they made a slave out of Pat in the South, right?. I got to stay in the hospital section of the chain gang. It wasn't too bad. While I was in Atlanta, a Cuban exile/dissident came to interview me in jail. I had no idea how he ever got my name or whereabouts. Caroline Kennedy's Secret Service agents interviewed me in the South. One of the agent's name was Williams-- The Will. He asked me if I had an interest in learning how to use explosives. I told him that my father's family had twenty-five to thirty years active during WWII and five major campaigns and that I wanted to try a writing career and a career as a political, psychological, religious and dissident writer. I stopped writing Carol, but I ended up giving in on everything over women-- single, celibate, exile, and the underworld purview. You name it. I started at the very bottom with the president's daughter, right? To develop my "Biomythography," I want to model my life after young Euclid's, St. Augustine's, and Dante's. Euclid wrote the <u>Elements</u>, and <u>Data</u>. So I will write about the scientific elements and data of my life. Augustine moved from a dissolute life to an effort to systematize Christian theology in Western Civilization. In contrast, I want to de-systematize theology in the image of entropy. The theme from Dante is "purgation to purification." To develop it, I like a theme from Yeats, "'Whatever has flamed upon the night man's own resinous heart had fed.'" I have experienced everything intellectually from the holocaust to metaphysical love, right? The "Biomythography" will move through this from "purgation to purification." My position is <u>The Divine Comedy</u> is an autobiography. Dante lost a war in Italy, his Vietnam.

Advice to Vance on Women, Life, and Culture

Whatever happened to your first wife? Was her name Sue? I liked her all right. She had the brains in the Opperman family. I heard from Theresa Lippert that you were on wife number three. Vance, what you want to tell your wives and children is that the name Opperman is not just a name it is a manifesto, an ideology, philosophy or theology. Everyone one of your families, actions, from writing to breast feeding, has metaphysical dimensions. At West Publishing, they should try to **refound world culture** at all levels--legal and psychological, for example. What do you know about the project evolution the Irish have with the South and the Semitic peoples? Are your children going to come up to your level?

A friend of mine from Benilde High School, Mark McGee, published with David Wilson a psychology textbook out of West Publishing. It sold well enough. Mark has written twelve books on relationships and many others. How many books have you published? I will contact Howie Kaibel. Also, win the toughest battle of all, the battle within Vance Opperman. Is there a big division at West on women's books? Another girl I went with was **Nancy Schoomer**. She grew up out on Lake Minnetonka. She had a blue collar father who paid for her violin lessons. She had a fight with him during college, and he pulled out a gun. Later, she met me and we went to Lakewood Cemetery for our date. I wanted her to experience a taste of love and death in the autumn. We broke up after awhile. She went to graduate school at the university in English. I wonder what ever happened to her. I recommend to you Andre Malraux an existentialist, socialist, Gaullist writer. He was a political activist like yourself. Have you ever read his book, <u>Man's Fate</u>? I gave it to my military/attorney dad when I was in high school at Washburn for a gift. The theme is the defining experience in a man's life is a taste of death. I tasted it with Theresa and Vance as an undergraduate at the University of Minnesota. I am with the Oblates of St. Benedict at St. Benedict's College, St. Joseph Minnesota, St. John's University, Collegeville, and Harvard. The shortcoming of secular education is that it leaves a "blank in your knowledge" about Catholicism. The Oblates fill this blank. It was wonderful hearing from you. Focus on Vance, God Bless, Patrick A. O'Dougherty, Ph.D.

See inclosure from the Paulist Press on other movements on campus.

May 28, 1991

Brother Patrick A. O'Dougherty
Hellenist America Catholic Ecumenical Graduate School
Riverside Plaza Apt. M3410
1615 South Fourth Street
Minneapolis, MN.: 55454

Robert W. Jacob
Coordinating Board For Higher Education
101 Adams Street
Jefferson City, Missouri: 65101-3059

Sir:

I am a lay brother with the Benedictines--a Benedictine Oblate. I have an interest in starting an Ecumenical Graduate School in the State of Missouri. I have published two books in American Church History: <u>An Existential Approach to American History and Walden III: A Catholic America</u>. Recently, I got the form-- Application for Recognition of Exemption. This form comes Under Section 501(c)(3) of the Internal Revenue Code. It contains Form 1023 and Form 872-C. On page one on Instructions for Form 1023 it says under number, 2, there are Organizations that are not required to file Form 1023. I think that I fall under this provision because I am a Catholic lay brother and because the gross receipts of my school will not be more that $5,000. I do not feel it necessary to receive and IRS determination letter.

IRISH PSYCHOLOGY/IRISH PSYCHIATRY: THE TIRESIAS COMPLEX

Here is a copy of these instructions and a copy of my Oblate form. I think that the Catholics should have a correspondence ecumenical graduate school.

Hope to hear from you soon,

Brother Patrick A. O'Dougherty

Dr. Terry Dosh
4124 Harriet Ave. So.
Minneapolis, Minnesota 55409

Heads on the Guillotine

Terry,   4/28/94

Where do Christ and the Blessed Mother fit in your restructuring of the Church? You didn't mention either of them by name in your discussion at Newman on Wednesday. I back the idea of a Catholic Constitutional Convention, but how would you relate it to the Gospels?

I favor dialogue rather than monologue with the one party liberal mentality at Newman. I hope there is dialogue in the Catholic Constitutional Convention. Buckley, Simon, Grace and Novak are formidable thinkers and Newman Center ignores them. I think that a good approach to Social Justice issues is the free enterprise approach. It is not considered. The Church should focus more on the producers in the world. Maybe the reason that a business is not making money is due to sin. This is America. We have more than one party.

My basic approach to Church history and Church restructuring is scientific. I have invented a new field--Cathecosophy. It is not the whole truth, but a very good approach to Catholic truth. Church history should be scientific and behavioral. The title of my book would be, The NonEuclidean Church: where faith and science intersect. This is the way the Church and Church history are going. Biblical personalities have a scientific dimension. This is an area where faith and science crisscross. I am enclosing a copy of my MMPI as an example of a scientific approach to a lay Brother's personality. Scientific psychology, economics, and the physical sciences should be a basis of theology and Catholicism. Idealism is also an approach to science. It offers a framework for understanding mystery. Catholicism over emphasizes realism. Did you know that Aristotle was a homosexual?

Freedom of enterprise is a basic Christian and human right.

217

IRISH PSYCHOLOGY/IRISH PSYCHIATRY: THE TIRESIAS COMPLEX

Why don't they present the Catholic home school movement at Newman?

Why do they leave the Blessed Mother out of the liturgies at Newman?

Why aren't the needs or ethnicity of the Irish or the Germans addressed at Newman?

Brother Patrick O'Dougherty, Ph.D.

10/6/94

Patrick A. O'Dougherty, Ph.D.
Riverside Plaza M3410
1615 So. Fourth Street
Minneapolis, Minnesota 55454

Judge Lance Allen Ito
Superior Court Magistrate
Los Angeles County
Los Angeles County Courthouse
111 North Hill St.
Los Angeles CA. 90012

Excellency:

I am not for an alleged woman killer.   They should not allow the defense to gang up on the woman attorney. I think that O.J. Simpson can really hurt Afro-American people's chances. I don't feel comfortable with O.J. I am remembering you in my prayers.

Patrick A. O'Dougherty, Ph.D.

Patrick O'Dougherty, Ph.D. 12/26/94 3:15p.m.

Pat Miles and Paul Majors
Kare Television-Gannett Broadcasting
8811 Olson Memorial Gldn. Vly
Fax: 546-8606
Below Average I.Q.s in the Windsor Family
Princess Diana's aptitude scores were so low that they were not going to let her graduate from high school. This means that she has an I.Q. in the 85-95 range.

218

IRISH PSYCHOLOGY/IRISH PSYCHIATRY: THE TIRESIAS COMPLEX

My mother, Patricia O'Dougherty, feels that Charles and his brothers and sister are very average.

Remember the Hindenburg
The Energy Crisis: The Big Lie

I have an interest in having Kare TV do a story on the Hindenburg Blimp crash. This was the reason that the automobile industry was given for not using hydrogen fuel in automobiles. There is no scientific proof for any fuel shortage. Hydrogen is the most common element in the universe and the most efficient fuel. This is the big lie that the world's public is being given. The world is being gassed, right?

In Minnesota, we lost the world's largest deposit of high grade iron ore in thirty years. My uncle is a Catholic priest in Arizona retired in a small copper mining town. The same event is happening there. America is going to lose its copper. The Japanese are getting our oil out of Alaska. This going to hurt everybody in the world, for example, the Portuguese, the Latvians and the English. There is almost no democracy in nature. Remember the Mesabi Iron Range. Remember the Hindenburg.

There is no scientific proof for any energy crisis. Call Dr. Mike Franey, a physicist at the University of Minnesota, for confirmation on this idea. Millions of people are getting cut down here, do you think this is some type of game? The woman wins,

Patrick O'Dougherty, Ph.D.

10/28/94

Terry and Millie,

I like William Buckley and Thomas Merton. There is very little democracy in nature. John Harris thinks "nature is a hierarchy, for example, scavengers, grazing animals, and predators." Nature is a terrible workings. Nature is a beautiful workings.

I like Royalism/Toryism, also I joined in the radical/socialist movement on campus: double trouble, double bind.

Royalism and radicalism are both major themes in Catholic history. Was Cardinal Newman was a royalist/radical?
Patrick O'Dougherty, Ph.D.

Patrick A. O'Dougherty, Ph.D. 12/28/94

President Mary Robinson

IRISH PSYCHOLOGY/IRISH PSYCHIATRY: THE TIRESIAS COMPLEX

Office of the President
'Aras An Uachtarain,
Phoenix Park
Dublin 8, Ireland

Below Average I.Q.s in the Windsor Family

Princess Diana's aptitude scores were so low that they were not going to let her graduate from high school. This means that she has an I.Q. in the 85-95 range. My mother, Patricia O'Dougherty, feels that Charles and his brothers and sister are very average. We want a new royal family in England.

Remember the Hindenburg
The Energy Crisis: The Big Lie

I want the Irish to do a story on the Hindenburg Blimp crash. This is the reason the American government gives for not using hydrogen fuel in automobiles. There is little scientific proof for any fuel shortage. Hydrogen is the most common element in the universe and the most efficient fuel. This is the big lie that the world's public is being given. Hydrogen is the most common element in Ireland.

In Minnesota, we lost the world's largest deposit of high grade iron ore in **thirty years.** My uncle is a Catholic priest in Arizona retired in a small copper mining town. The same action is happening there. America is going to lose its copper. The Japanese are getting our oil out of Alaska. This going to hurt everybody in the world: the Portuguese, the Latvians, and the Irish. Remember the Mesabi Iron Range. Remember the Hindenburg. There is no scientific proof for any energy crisis. The Arabs will lose their oil in about 50 years, and they have little else there. They should change over to hydrogen slowly.

The Catholic intellectual outlook is completely unacceptable to most Americans. Most of the people who have opposed the Church in Western Civilization have ended losers.

Patrick A. O'Dougherty, Ph.D.

Patrick A. O'Dougherty, History of Science, Ph.D.

Chairman Alexander Trotman
Ford Motor Company: The American Road
Dearborn, Michigan 48121

1/14/95

Sir:

Remember the Hindenburg
The Energy Crisis: The Big Lie

220

IRISH PSYCHOLOGY/IRISH PSYCHIATRY: THE TIRESIAS COMPLEX

I have an interest in having Ford Motor Company do research on the Hindenburg Blimp crash. This is the reason the American government gives for not using hydrogen fuel in automobiles. There is little scientific proof for any fuel shortage. Hydrogen is the most common element in the universe and the most efficient fuel. This is the big lie that the world's public is being given. Hydrogen is the most common element in the universe.

In Minnesota, we lost the world's largest deposit of high grade iron ore in **thirty years**. My uncle lives in Arizona retired in a small copper mining town. The same event is happening there. America is going to lose its copper. The Japanese are getting our oil out of Alaska. This going to hurt everybody in the world: the Germans, the Portuguese, the Latvians, and the Irish. Remember the Mesabi Iron Range. Remember the Hindenburg. There is no scientific proof for any energy crisis. The Arabs will lose their oil in about 50 years, and they have little else there. They should change over to hydrogen slowly.

Patrick A. O'Dougherty, Ph.D.

Terry Dosh,   1/19/95

We are all published in our family. Tell Millie, Paul and Martin to send their publications to me after they break out into print. It would be more objective that way. Have you ever read Thomas Mann's, <u>Buddenbrooks</u>? It is about the decline and fall of a German family, like the Windsor's, after several generations. He is my favorite German author. He got a Nobel Prize. I think that most of his books will stand the test of time. They are classics.

Patrick A. O'Dougherty, Ph.D.

5/5/95
Cinco de Mayo

Patrick A. O'Dougherty, Ph.D., Purple Heart Veteran of Urban America.

Dr. Patrick A. O'Dougherty has spent twenty-five years in the combat zone: Atlanta, South Minneapolis, Inner City St. Paul. He has experienced several murder attempts, four plus robberies and countless gang rivalry situations. He has had thirteen plus years at a business front door in the black community. He ran a political election out of the black community. He is founder of the field of genetic warfare. He is a member of the "cultural war" movement of America.

221

IRISH PSYCHOLOGY/IRISH PSYCHIATRY: THE TIRESIAS COMPLEX

Minnesota Atheists
P.O. Box 6261
Minneapolis, Minnesota 55406

Atheist Letters: The Communications Corner
The Death of the Dream: The Suicide of the Human Race
12/9/95

Life is a gift. God is "self-gift." The universe is God's self-gift or "God's dream," like Martin Luther King's dream. One goal of the Catholic is to come up to God's level of self-gift, the universe, intellectually. The word, "God," relates etymologically ("truth plus word plus reason") to the word "good."

(Source: John O'Dougherty) Charity defines the good and the common good. Animals do not live for the common good. The mark of the Catholic is charity which is gift and self-gift. The mystery of the cross is sacrificial love or self-gift. Without sacrifice there is little excellence. Evolution turns into devolution. The human race sinks back into the goo or the material or "manure." Atheism is materialism or a "manure philosophy." I decline to become a member of Minnesota Atheists. I decline to accept the death of the dream. I decline to accept the suicide of the human race.

Patrick A. O'Dougherty, Ph.D., Minnesota and Harvard

John Herlick: 4/1/96
The Green Revolution Worldwide

Happy fiftieth birthday! For your birthday I am sending you some information on the Constitution for the Federation of Earth and the World Parliament. I am for the Green Party, Green Caucus internationally. Petra Kelly founded the Green Party in Germany. Does the Republican Party contain a Green Caucus? I am also sending you some information on silver mercury fillings which Europeans banned.
God Bless,

Patrick A. O'Dougherty, Ph.D.

"'Politics is the shadow big business casts on society.'" John Dewey

Dear Citizen: 7/4/96
Jefferson and Jesus

Because Jefferson was a leader in the movement for religious freedom, many people of his time labelled him an atheist. Jefferson did, however, spend much of his time reflecting on the life of Jesus of Nazareth. His little book on his version of

222

the Gospels looks at his idea of the true sayings of Jesus. This book, <u>The Life and Morals of Jesus of Nazareth</u>, the Fifty-Seventh Congress published and each member received a copy.[198]

What prompted Jefferson to write this book? He felt that many tricks had been played on the Biblical text. For example, in the New Testament Jefferson argued there is an sign that aspects of the text flowed from a gifted man. Yet other lines in the text were mediocre. Jefferson tries to separate the jewels from the average.[199]

Jefferson referred to himself as a Christian; but, he had opposition "to the corruptions of Christianity." Jefferson ascribed to Jesus every "human" virtue. The passage "who ever blasphemes against the Holy Spirit never has forgiveness but is guilty of an eternal sin" horrified Jefferson. He favors Mark who says, "Who ever believes and is baptized will be saved, but who ever does not believe will be damned." Jefferson also found that many of the teachings of the world's religions were similar.[200]

Senator Linda Berglin
61st District DFL
G-10 State Capitol
St. Paul, MN 55155

Senator Linda Berglin:

2/23/97

I'm in your district. Here are some of my projects. I want to set up a time at the State Capitol to go over them with you. I ran for mayor of Minneapolis in the DFL caucus the first year that you ran for office. I am a leading theoretician in the Green Party.

MAJOR ENVIRONMENTAL ENERGY PROJECTS
THE HYDROGEN CONVERSION PROJECT

Experts suggest that the biggest problem in the economy of the State of Minnesota is that it has to import 98 percent of its fuel. There is an easy alternative to this: hydrogen fuel. If tapped correctly, there is enough energy in the hydrogen in a glass of water to power St. Paul for a winter. Hydrogen fuel is the most common, most efficient, and a very powerful and a very clean fuel. Hydrogen composes ninety percent of the universe. It can power automobiles, planes, even space missions. The reason we don't use hydrogen in automobiles is because of the Hindenburg Blimp crash. The claim is that it is too unstable as a fuel. It is more stable than fossil fuels. Most countries, for example, Israel, China and Poland do not have fossil fuels. It took millions of years to build up the fossil fuel reserves in nature. It would be a mistake to deplete these reserves rapidly. War would be the outcome. In Japan people have to wear gas masks to protect themselves from the pollution. In this country Lake Erie is dead. The byproduct of hydrogen is water vapor which has irrigation potential. The main economic problems in most North American states are fuel and gas importation. A spinoff of this project would be a hydrogen powered Northern Space Center. Ireland, England, the Scandinavian countries and Germany are changing over to hydrogen fuel. Saab and Mercedes are changing over to hydrogen fuel. Electric vehicles are another important alternative. France, Japan, and England are rapidly introducing them. We need a model hydrogen fueled city in Minnesota. Hydrogen is a continuously recyclable fuel.

THE ALTERNATIVES TO PETROCHEMICALS PROJECT
by Patrick A. O'Dougherty, Ph.D., and Terry Gips, IASA

A spin off the hydrogen conversion project is the alternatives to petrochemicals project. Petrochemicals are in many products, for example, plastics and aerosols. What are biodegradable functional equivalents of them? For example, the University of Minnesota Medical School needs a house cleaning. It is on the bluffs of the Mississippi River on the campus of the University of Minnesota. Ironically, the Mississippi River has become the river of death or "Cancer Alley" because of many oil refineries and petrochemical plants which dot its banks. There petrochemical plants account for one quarter of U.S. petrochemical manufacturing. Most of these plants concentrate along a seventy-five mile strip from Baton Rouge to New Orleans, LA called "Cancer Alley." Ironically, along this river there are hundreds of industrial facilities producing toxic pollutants. In the medical school petrochemicals are everywhere, for example, in plastics, fuels, skin care products, cleaning products, rubbers, paint, synthetic fibers, and bandages. Many products that do not contain petrochemicals come from them, for example, computer circuit boards.

Living Dangerously

Petrochemicals and oil manufacturing precipitate climate change, affect heathy living, ecosystem destruction, game and fish poisoning, and animal and human rights violations.

Vital Data

The U.S. released nearly one billion pounds of petrochemical toxins into the U.S. air, water and soil in 1993. 90 million barrels or oil leak into the U.S. each year. In 1990, vehicle exhaust deposited nearly 8 million tons of volatile organic compounds into U.S. air creating smog. CO_2 from fossil fuel burning creates the largest source of greenhouse gases by humans. Petrochemical pesticides and fertilizers have contaminated the ground water, as well as the workers and wildlife. There are nearly 24,000 different types of consumer products in the U.S. such as aerosols and perfumes to name a few which release nearly 1 million tons of volatile organic compounds into the air.

Health and Petrochemicals

High-level exposures create headaches, dizziness, blindness, asthma and death among other side effects. Low-level long-term exposures produce nausea, cancer, lung damage, birth defects, sterility, and liver and kidney damage to name just a few problems. For example, hospital personnel handling these petrochemical products are more likely to develop respiratory or intestinal cancers. Also, seven out of the top eight U.S. baby foods contain one or more pesticides containing DDT. This is probably true for medical school or hospital baby foods.

A New Beginning

The Petrochemical Alternatives Campaign is growing by people who want to provide information that will help lessen the widespread use of petrochemicals which are often toxic.

Praxis

You can use organic foods, petrochemical free products and pollution free forms of transportation. You can become an activist and revolutionist in the Petrochemical Alternatives Campaign. You can vote for responsible change in the field. You can speak your mind, and you can learn about alternatives to petrochemicals.

THE BIOMASS CONVERSION PROJECT AND THE HYDROGEN CONVERSION PROJECT

by Patrick A. O'Dougherty, Ph.D.

Biomass, for example, trash converts to hydrogen fuel. Convert your trash to biomass!

THE NUCLEAR WASTE CONVERSION PROJECT

by Patrick A. O'Dougherty, Ph.D.

Large amounts of research, effort, planning and business investment have gone into nuclear research. The problem now is to get scientists to figure out how to make nuclear waste benign, recyclable and biodegradable.

Reference on Alternative to Petrochemicals: A work in progress by Terry Gips from the Petrochemicals Alternatives Campaign. The International Alliance for Sustainable Agriculture, Newman Center, 1701 University Ave SE., Mpls., MN, 55414.

You are doing a great job,

IRISH PSYCHOLOGY/IRISH PSYCHIATRY: THE TIRESIAS COMPLEX

Patrick A. O'Dougherty, Ph.D., Minneapolis and Harvard

Reference on Alternatives to Petrochemicals: A brochure in the works Terry Gips
from the Petrochemicals Alternatives Campaign. The International Alliance for
Sustainable Agriculture, Newman Center, 1701 University Ave. SE., Mpls., MN.
55414.

Senator Linda Berglin
61st District DFL
G-10 State Capitol
St. Paul, MN 55155

Senator Linda Berglin:

2/23/97

Here are some of the issues that I am working on as a theoretician for the Green
Party.

THE GREEN PARTY AGENDA

THE COMMUNAL PROPERTY AMENDMENT
The American Constitution: Through a Green Glass Darkly
   St. Paul's vision was "through a glass darkly." Greens should see the
American Constitution through a green glass darkly. For example, the Fifth
Amendment to the Constitution protects the right of private property: "nor shall
private property be taken for public use without just compensation." A Green
Communal Property Amendment to the United States Constitution is necessary to
protect for seven generations the nearly twenty-five percent of America that is
communal property, for example, the coasts, the mountain ranges, the parks, the
rivers, and your backyard. Winona LaDuke supports a Communal Property
Amendment. The problem is dense, like a forest. Definition is one key to the
problem. Communal rights and responsibilities are another approach. In Tennessee,
Kentucky, Spain and Germany residents often own their homes, but the rights to the
resources under their homes are held by the community or the government. The
stakes are high. For example, Minnesotans lost the world's largest deposit of high
grade iron ore during the WWI and WWII years due to a lack of communal or
common good protection of this precious resource. Good ecology is good green dark
constitutional theology. Every amendment in the American Constitution is a Green
Amendment.

A GREEN REVOLUTION/GREEN PARTY BOOK

IRISH PSYCHOLOGY/IRISH PSYCHIATRY: THE TIRESIAS COMPLEX

Dr. Patrick O'Dougherty has written a book for the Green Party addressing the Green Revolution. It is <u>St. Patrick, The Green Revolution, and the Hydrogen Conversion Project</u>. A few of the major topics are St. Patrick, the interrelatedness of all species, a critique of the Shallow versus Deep Ecology dualism, Green Revolution issues, a personalist/intuitionist philosophy, hydrogen conversion, electric vehicles, biomass conversion, the Green Party, the movement for a Constitution for the Federation of the Earth, the movement for a World Parliament, the Ecozoic Age. It includes a twenty-nine page sustainable agriculture data base.

THE SEVENTH GENERATION AMENDMENT
The Greens are working on an amendment to protect much of our civilization for seven generations.

THE PROPORTIONAL REPRESENTATION AMENDMENT
Basically, proportional representation means if you get seven percent of the vote, you get seven percent of the seats in the congress. Most of the other countries in Western Civilization have this approach and New Zealand just changed its lower house to proportional representation.

THE FEDERAL RESERVE BOARD: REGAINING DEMOCRATIC CONTROL
The United States Constitution says that only the Congress has the power to issue or regulate money. As a <u>wartime measure</u> during WWI the U.S. Government created the Federal Reserve Board to regulate the currency. The people lost democratic control over its currency by the creation of the Federal Reserve Board. The people want to regain democratic say over currency regulation.

THE STOCK MARKET: A DEMOCRACY?
Why isn't the stock market and stock market investment run by the people and for the people? Why isn't it run through our democracy and democratic institutions? The stock market should be a democracy.
PARITY IN TAXATION
The big problem in Minnesota taxation is the this state pays out to the federal government much more money than it gets back, and it pays out much more money than it gets back in comparison to other states, for example, Texas and California. How can the State of Minnesota address parity in taxation? Patrick A. O'Dougherty, Ph.D., the Green Party

Riverside Plaza M3410 1615 So. 4th. St. Mpls., MN 55454
(612) 339-1748

BIBLIOGRAPHY

American Council of Learned Societies. editors. Dictionary of Scientific Biography. New York: Charles Scribners Sons, Publishers, Vol. II, 1981.

American Psychiatric Association. Diagnostic and Statistical Manual of Mental Disorders. Washington: 1994.

Bible.

Brady, Ciaran. editor. Interpreting Irish History: The Debate on Historical Revisionism 1938-1994 Dublin: Irish Academic Press, 1994.

Citizens Commission on Human Rights. Los Angeles: Citizens Commission on Human Rights, 1995.

Conant, Brother Gregory, OSB. St. Benedict's Abbey. Still River, Massachusetts.

Conroy, Pat. The Great Santini. Boston: Houghton Mifflin, 1976.

Day, Dorothy. The Catholic Worker. Vol. LVIII, No. 7, October-November, 1991.

Evans, Ivor H. Brewer's Dictionary of Phrase and Fable. New York: Harper & Row, Publishers, 1981.

Freeman, Father Hilary. former professor of logic, St. Catherine's College, St. Paul, Minnesota. St. Thomas Aquinas Priory. River Forest, Illinois.

Green Party of Minnesota. Arise Book Store & Resource Center. 2441 Lyndale Avenue So. Minneapolis.

Guralnik, David B. editor. Webster's New World Dictionary of the American Language. New World: Simon and Schuster, 1980.

Howaton, M.C. editor. The Oxford Companion to Classical Literature. Springfield: Merriam Webster, Publisher, 1995.

IRISH PSYCHOLOGY/IRISH PSYCHIATRY: THE TIRESIAS COMPLEX

Jungian Personality Test. Personality Archetypes. unknown publisher.

Kuiper, Kathleen. editor. Merriam-Webster's Encyclopedia of Literature Springfield: Merriam-Webster, Publishers, 1995.

Lassus, Arnaud de. Apropos. October, 1989.

Manes, Christopher. Green Rage: Radical Environmentalism and the Unmaking of Civilization Boston: Little, Brown and Company, 1990.

McCarthy, Father Charlie. The Catholic Worker. Vol. LVIII, No. 7.

McGee, Mark. psychologist. Aurora, CO. pseudonym. Faraway I. Till. Constructive Replication of Thoreau's Experiment. Aurora, CO: Mentor Institute, 1990.

MMPI1-2: The Minnesota Report Adult Clinical System Interpretive Report, 15-April, 1995.

Mitchell, Stephen. The Gospel According to Jesus. New York: Harper Collins, Publishers, 1991.

Noble, David. history professor. University of Minnesota. Minneapolis, Minnesota.

O'Dochartaigh, Fionbarra. ulster's white negroes: from civil rights to insurrection. Edinburgh: AK Press, 1994.

O'Dougherty, John. an uncle. Minneapolis, Minnesota

Pegis, Anton C. editor. Introduction to Saint Thomas Aquinas. New York: The Modern Library, Random House, 1948

Rodgers, Martin. Catholic Worker. Vol. LVIII, No. 6, September 1991.

Skinner, B. F. Beyond Freedom and Human Dignity. New York: Knopf, 1971.

Strong, Edward K. Strong Vocational Interest Test-Men. Stanford University, 1963.

IRISH PSYCHOLOGY/IRISH PSYCHIATRY: THE TIRESIAS COMPLEX

Swimme Brian & Thomas Berry. The Universe Story. New York:
    HarperSanFrancisco A Division of Harper Collins Publishers, 1992.

Travasio, Joseph. Minneapolis, Minnesota.

Wagner, Sister Mary Anthony.   Oblates of St. Benedict.   College of
    St. Benedict. St. Joseph, Minnesota.

Webster's New Universal UnAbridged Dictionary.   New   York: Barnes
    & Noble Books, 1994.

Welsh,   Bishop   Lawrence.   Archdiocese   of   St.   Paul.   St.   Paul,
    Minnesota.

Wojtyla, Karol.   Pope John Paul II.   Love and Responsibility.   New
    York: Farrar, Straus, Girous, 1981.

ABBREVIATED INDEX

IRISH PSYCHOLOGY/IRISH PSYCHIATRY: THE TIRESIAS COMPLEX

IRISH PSYCHOLOGY/IRISH PSYCHIATRY: THE TIRESIAS COMPLEX

Footnotes for Irish Psychology/Irish Psychiatry

1. Arnaud de Lassus, "Religious Liberty," <u>Apropos</u>, October, 1989: 6.
2. Brother Gregory Conant, OSB, St. Benedict's Abbey, 252 Still River Road, Still River Massachusetts 01461.
3. Lassus, <u>op. cit.</u>, 6.
4. <u>Ibid</u>., 7. Martin Luther King, Roisin McAliskey and Ghandi are the best examples of the idea that physical freedom is a necessary part of freedom. They were all imprisoned for their beliefs.
5. <u>Ibid</u>., 10-11.
6. <u>Ibid</u>., 8-9.
7. <u>Ibid</u>., 12-13.
8. <u>Ibid</u>., 12.
9. <u>Ibid</u>., 12-13.
10. <u>Ibid</u>.
11. <u>Ibid</u>.
12. <u>Ibid</u>., 7.
13. <u>Ibid</u>., 12.
14. <u>Ibid</u>.
15. <u>Ibid</u>., 20.
16. <u>Ibid</u>., 18.
17. <u>Ibid</u>., 21.
18. <u>Ibid</u>., 24.
19. <u>Ibid</u>., 12-13.
20. <u>Ibid</u>., 28.
21. <u>Ibid</u>., 12.
22. <u>Ibid</u>., 28.
23. <u>Ibid</u>.
24. <u>Ibid</u>., 9. Freedom is essentially anti-colonialism. Its fulfillment and praxes are best found and developed in anti-colonial struggles and women's and minorities struggles.
25. B.F. Skinner, <u>Beyond Freedom and Human Dignity</u> (New York: Knopf, 1971.
26. Lassus, <u>op. cit.</u>, 16-17.
27. <u>Ibid</u>.
28. <u>Ibid</u>., 23.
29. <u>Ibid</u>., 24.
30. <u>Ibid</u>., 24-25.
31. <u>Ibid</u>., 25.
32. <u>Ibid</u>., 21.
33. <u>Ibid</u>., 23.
34. <u>Ibid</u>.
35. <u>Ibid</u>., 14-15.

36. Ibid., 22.
37. Ibid., 29.
38. Sister Mary Anthony Wagner, St. Benedict's Monastery, St. Joseph, Minnesota.
39. John O'Dougherty, an uncle, used this expression.
40. Fionbarra O'Dochartaigh, ulster's white negroes: from civil rights to insurrection (Edinburgh: AK Press, 1994).
41. Ibid., cover.
42. Ibid., 119-120.
43. David B. Guralnik, editor, "personalism," Webster's New World Dictionary of the American Language, Second College Edition (New York: Simon and Schuster, 1980), 1062
44. Pope John Paul II, Karol Wojtyla, Love and Responsibility (New York: Farrar, Straus, Girous, 1981), 23.
45. Wojtyla, op. cit., 72-73.
46. Ibid., 74-96.
47. Guralnik, op. cit., "language," 792.
48. Ibid., "mathematics," 875.
49. Father Hilary Freeman maintains that mathematicians cannot agree on how to define their field.
50. Gurlanik, op. cit., "methodology," 895.
51. American Council of Learned Societies, editors, "Brouwer, Luitzen Egbertus Jan," Dictionary of Scientific Biography (New York: Charles ScribnersSons, Publishers), Vol. II. 512.
52. Ibid.
53. Ibid.
54. Dorothy Day, "A Season Crowned in Glory," The Catholic Worker, Vol. LVIII, No. 7, October-November, 1991.
55. Ibid. p. 6.
56. Ibid.
57. Ibid.
58. Father Charlie McCarthy, "St. Therese and the Little Way," The Catholic Worker, Vol. LVIII, No. 7, pp. 1 & 7.
59. Ibid.
60. Ibid.
61. Martin Rodgers, "I Recognize God in You," Catholic Worker, Vol. LVIII, No. 6, September 1991, pp. 1 & 5.
62. Ibid.
63. Ibid.
64. Guralnik, op. cit., "personalism," 1062.
65. Anton C. Pegis, editor, Introduction to Saint Thomas Aquinas (New York: The Modern Library, Random House, 1948), 5-6.
66. Ibid.

67.   Ibid.
68.   Ibid., 114-115.
69.   The writer and Joe Travasio, a leftwing thinker.
70.   "Words are Deeds," this idea was discussed in a lecture at a conference on Wittgenstein sponsored by the philosophy department at the University of Minnesota in the Fall of 1996.
71.   Christopher   Manes,   "activism,"   Green   Rage: Radical Environmentalism and the Unmaking of Civilization (Boston: Little, Brown and Company, 1990), 94.
72.   Ibid.
73.   Ibid., "Adat," 172-173, 185, 239.
74.   Ibid., 122-123, 172-173, 239-240.
75.   Ibid., 19.
76.   Ibid., "anti-drug bill," 12.
77.   Ibid., 19-20, 135.
78.   Ibid., "antinature," 41-43.
79.   Ibid., 119.
80.   Ibid., 153-154.
81.   Ibid., "tree spiking," 10.
82.   Ibid., 227.
83.   Patrick A. O'Dougherty, the writer, is sponsoring the Communal Property Rights Amendment in the Green Party.
84.   Ibid., "confrontation," 3-7.
85.   Ibid., 167.
86.   Ibid., "cracking," 6.
87.   Ibid., 168.
88.   Ibid., 228.
89.   Thoreau thought life is "counterfriction."
90.   Lost reference from a television special.
91.   Brian Swimme & Thomas Berry, The Universe Story (New York: HarperSanFrancisco A Division of Harper Collins Publisher, 1992), 253-258.
92.   Manes, op. cit., 38.
93.   Ibid., 229.
94.   Ibid., 10.
95.   Michel Foucault thinks "discontinuity is the stuff of history" 42.
96.   Ibid., 161.
97.   Ibid., 50.
98.   Ibid., 235.
99.   Ibid., 4-69.
100. Ibid., 248.
101. Ibid., 36.

102. _Ibid._, 24.
103. _Ibid._, 225.
104. _Ibid._, 93.
105. _Ibid._, 160.
106. _Ibid._, 229.
107. _Ibid._, 109.
108. _Ibid._, 33.
109. _Ibid._, 172-173, 185, 239.
110. _Ibid._, 91. John N. Harris, psychologist also mentioned this topic.
111. _Ibid._, 235.
112. _Ibid._, 22.
113. _Ibid._, 62.
114. _Ibid._, 152.
115. _Ibid._ 71.
116. _Ibid._, 60.
117. _Ibid._, 174.
118. _Ibid._, 8-9.
119. _Ibid._, 90.
120. _Ibid._, 21, 155.
121. _Ibid._ 41.
122. _Ibid._, 151.
123. _Ibid._, 111.
124. _Ibid._ 158.
125. _Ibid._, 58.
126. _Ibid._, 98.
127. _Ibid._, 50.
128. _Ibid._.
129. _Ibid._, 106.
130. _Ibid._, 33.
131. _Ibid._, 24. This counters the idea of a land ethic.
132. _Ibid._, 34.
133. _Ibid._, 82.
134. _Ibid._, 121.
135. _Ibid._, 155.
136. _Ibid._
137. Citizens Commission on Human Rights, "Psychiatry's Betrayal," (Los Angeles: Citizens Commission on Human Rights, 1995), 5.
138. _Ibid._, 231.
139. _Ibid._, 48.
140. _Ibid._, 20.
141. _Ibid._, 156.
142. _Ibid._, 203.

IRISH PSYCHOLOGY/IRISH PSYCHIATRY: THE TIRESIAS COMPLEX

143. Citizens Commission on Human Rights, op. cit., "Psychiatry and South Africa," 16-19.
144. Ibid., "How Psychiatry Lit the Racial Fires," 4-7.
145. Ibid., "Psychiatric Oppression of African Americans," 8-11.
146. Ibid., 133.
147. Ibid.
148. Ibid.,46.
149. Ibid., 79.
150. Ibid., 26.
151. Ibid., 81.
152. Ibid., 26.
153. Ibid., 236.
154. Mark McGee, psychologist, has tried extensively as a writer to recapitulate the Thoreau experiment in his life and writings. Irish Psychology/Irish Psychiatry is essentially a recapitulation of the Thoreau experiment in an urban Walden. The writer's life as a psyche client is thus a recapitulation of the Thoreau experiment. Mark McGee writes under the pseudonym, Faraway I. Till. His work, Constructive Replication of Thoreau's Experiment, is published by Mentor Institute, Inc. P.O. Box 460562, Aurora, CO 80046-0562.
155. Ibid., 99.
156. Ibid.,236.
157. Christopher Manes, Green Rage: Radical Environmentalism and the Unmaking of Civilization (Boston: Little, Brown and Company), et passim. The librarian at the Minneapolis Public Library told me that because I had created a completely new dictionary, a radical psychology/psychiatry dictionary, out of the environmental terms in Green Rage this new dictionary is mine and I do not have to footnote every term in it. This footnote gives Christopher Manes credit et. passim for his input into my dictionary. However, the writer takes credit for the terms and definitions in this dictionary in miniature.
158. Green Party of Minnesota State Party Platform 1996, Green Party of Minnesota P. O. Box 582931 Minneapolis, MN 55458.
159. Bishop Lawrence Welsh, Archdiocese of St. Paul, mentioned this theme in his Lent retreat at the St. Paul Cathedral.
160. Rev. Francis Evans, editor, Prayers For All Occasions (New York: Catholic Book Publishing Co., 1987), 44.
161. MMPI1-2: The Minnesota Report Adult Clinical System Interpretive Report, 15-April, 1995.
162. Joyce Forsgren, coordinator, Day Treatment, Hennepin County Medical Center, Minneapolis, Minnesota.
163. Ibid.

IRISH PSYCHOLOGY/IRISH PSYCHIATRY: THE TIRESIAS COMPLEX

164. Ivor H. Evans, "Tiresias," Brewer's Dictionary of Phrase and Fable (New York: Harper & Row, Publishers, 1981), 1120.
165. M.C. Howaton, editor, "Teiresias," The Oxford Companion to Classical Literature, Second Edition (New York: Oxford University Press, 1989), 550
166. Ibid.
167. Kathleen Kuiper, editor, "Tiresias," Merriam-Webster's Encyclopedia of Literature (Springfield, Massachusetts: Merriam-Webster, Publishers, 1995), 1117.
168. These psalm selections came from Joan Chittister's, The Rule of Benedict: Insights for the Ages (New York: The Crossroad Publishing Company, 1992), 82.
169. Evans, op. cit., 150
170. Mark McGee, a friend and psychologist, gave me this poem as a token of our friendship.
171. Ciaran Brady, editor, Interpreting Irish History: The Debate on Historical Revisionism 1938-1994 (Dublin: Irish Academic Press, 1994), 198-199.
172. Ibid., 302-314.
173. Ibid., 210.
174. Ibid., 211-214.
175. Ibid., 214-215.
176. Ibid., 216-229.
177. Mark McGee, psychologist.
178. Brady, op. cit., 249-251.
179. Ibid., 302.
180. Ibid., 314.
181. Ibid.
182. Ibid., 315-317.
183. Ibid., 14.
184. Ibid., 314.
185. Ibid., 315.
186. Ibid., 315-317
187. Ibid.
188. Ibid.
189. Ibid., 317-323.
190. Webster's New Universal UnAbridged Dictionary, "cuckoo," (New York: Barnes & Noble Books, 1994), 351. Professor Richard Rudolph, Department of Austrian Studies, University of Minnesota made the observation in a lecture that the methodology of the field of history is like that of the cuckoo bird which borrows from the nests of other birds. I am indebted to him for this observation.
191. Roger Hallebuyck, a friend, St. Paul, Minnesota.

IRISH PSYCHOLOGY/IRISH PSYCHIATRY: THE TIRESIAS COMPLEX

192. John Noel Harris, psychologist, Hopkins, Minnesota.
193. Ibid.
194. Ibid.
195. David B. Guralnik, editor, "royalist," <u>Webster's New World Dictionary of the American Language</u> (New York: Simon & Schuster, 1980), 1242.
196. Ibid., "Tory," 1502.
197. Ibid., "radical," 1171.
198. Stephen Mitchell, <u>The Gospel According to Jesus</u> (New York: HarperCollins, Publishers, 1991), pp. 3-4.
199. Ibid. p. 5.
200. Ibid. pp. 5-9.

World Shakespeare Bibliography

James L. Harner, editor

Department of English   Texas A&M University   College Station, TX 77843-4227
409-845-3400   409-862-2292 (fax)   jlh5651@venus.tamu.edu

16 April 1996

Patrick A. O'Dougherty
Hellenist America Institute Publishing Co.
Riverside Plaza M3410
1615 S. Fourth St.
Minneapolis, MN 55454

Dear Dr. O'Dougherty,

    Many thanks for the loan of the enclosed copy of your <u>Shakespeare and Dream Work</u>. I've written up an entry for the 1995 Bibliography (and the <u>World Shakespeare Bibliography on CD-ROM 1900-Present</u>).

    Do keep me informed if you publish further books or articles on Shakespeare.

Yours sincerely,

James L. Harner

RUN DATE: 12/10/96
SORT: PUB/IMPRINT/TITLE
PO BOX 5000-0103
OLDSMAR, FL 34677-0103

HELLENIST AMER
REED REFERENCE PUBLISHING - BOWKER
BOOKS IN PRINT CHECKLIST

DEADLINE: 02/04/97
BOWKER CONTACT NAME: LESLIE FISHER
BOWKER CONTACT PHONE: 813-891-7908

TITLE:An Existential & Numerical Approach to American History   SUB-TTL:The American Revolution: A Case Study

ISBN:0-9626665-0-5   CONTRIBS:AU Patrick A. O'Dougherty   BINDING:Cloth Text   PRICE:$10.99   PUB-DATE:10/1996   PGS: 151

AUDIENCE:C   LCCN:96-094077   PUB-SYM:Hellenist Amer Co

TITLE:Life Culture Versus Death Culture & the Death of Literature

ISBN:0-9626665-3-X   CONTRIBS:AU Patrick A. O'Dougherty   BINDING:Library Binding   PRICE:$10.99   PUB-DATE:12/1995   PGS:50

AUDIENCE:C   PUB-SYM:Hellenist Amer Co

TITLE:Patrick's "Unfinished". An Intellectual History Counterpoint to Franz Schubert's Symphony No. 8 "Unfinished"   SUB-TTL:An Existential Approach to American History

ISBN:0-9626665-6-4   CONTRIBS:AU Patrick A. O'Dougherty   BINDING:Trade Paper   PRICE:$5.99   PUB-DATE:05/1996   PGS:60

ED-INFO:UBR   AUDIENCE:C   LCCN:96-094077   PUB-SYM:Hellenist Amer Co

TITLE:Personalism & Mathematics As Women's Personifestors   SUB-TTL:Women & the Fior (Which Is Irish for Truth) & the Creation of the Personalist Intuitionist School of Mathematics & Physics

ISBN:0-9626665-5-6   CONTRIBS:AU Patrick A. O'Dougherty   BINDING:Library Binding   PRICE:$20.99   PUB-DATE:10/1995   PGS:106

AUDIENCE:C   PUB-SYM:Hellenist Amer Co

RUN DATE: 12/10/96
SORT: PUB/IMPRINT/TITLE
PO BOX 6000-0103
OLDSMAR, FL 34677-0103

TITLE:Reinventing Physics   SUB-TTL:Logic & Physics, a Dialectical Approach to Physics

ISBN:0-9626665-2-1   CONTRIBS:AU Patrick A. O'Dougherty   BINDING:Library Binding   PRICE:$10.99   PUB-DATE:10/1993   PGS:144

AUDIENCE:C   PUB-SYM:Hellenist Amer Co

TITLE:Shaking up Shakespeare: His Dreamwork & Personality   SUB-TTL:Shakespeare: Dreamwork, Personality & Complexity

ISBN:0-9626665-4-8   CONTRIBS:AU, ED, Introduction by Patrick A. O'Dougherty   BINDING:Library Binding   PRICE:$19.99

PUB-DATE:11/1994   PGS:169   ILLUS:yes   AUDIENCE:C   PUB-SYM:Hellenist Amer Co

TITLE:St. Patrick, the Green Revolution, & the Hydrogen Conversion Project   SUB-TTL:Featuring the International Alliance for

Sustainable Agriculture Purple Database

ISBN:0-9626665-7-2   CONTRIBS:AU, ED, Introduction by Patrick A. O'Dougherty   BINDING:Library Binding   PRICE:$19.99

PUB-DATE:10/1996   PGS:116   ILLUS:yes   PUB-SYM:Hellenist Amer Co

TITLE:Walden III   SUB-TTL:A Catholic America

ISBN:0-9626665-1-3   CONTRIBS:AU Patrick O'Dougherty   BINDING:Library Binding   PRICE:$10.99   PUB-DATE:04/1991   PGS:92

AUDIENCE:C   LCCN:96-079986   PUB-SYM:Hellenist Amer Co

TOTAL NUMBER OF TITLES: 8

www.ingramcontent.com/pod-product-compliance
Lightning Source LLC
Chambersburg PA
CBHW031828170526
45157CB00001B/223